PLANTING
IN A
POST-WILD
WORLD

PLANTING
IN A
POST-WILD
WORLD

DESIGNING PLANT COMMUNITIES FOR RESILIENT LANDSCAPES

THOMAS RAINER AND CLAUDIA WEST

Timber Press
Portland, Oregon

CONTENTS

< White wood asters creep underneath Japanese beech ferns, densely covering the ground.

THOMAS

In the summer before my second grade year, we moved to just outside of Birmingham, Alabama. My family bought a house in a new development on the edge of town. The summer we moved in, there were half a dozen other houses on our street, with as many empty woodlots. But within a couple of years, the woodlots disappeared. New houses, new families, and new children filled in the street. With no more woodlots to explore, I turned my attention to the enormous forested tract of land that bordered our backyard. The steel company owned the land, and since they did not manage or tend it, I spent my weekends and summers romping through those woods with a pack of irreverent boys from the neighborhood. The Piedmont forest stretched several square miles in all directions and connected with even larger tracts of undeveloped land beyond. We spent our days building lean-tos and forts, evading enemies (usually younger sisters), foraging muscadines and dewberries, and exploring the outer boundaries of these seemingly endless wilds.

My earliest memory of wild plants was of the rich spaces they created. A tangled thicket of sparkleberry trees formed narrow paths which we moved through like rabbits; a massive southern red oak was our meeting spot; and perhaps most sacred of all was a grove of beeches whose canopy created a dome over a wide creek that flowed between two ridges. We would descend into that bowl in silence, entranced by the light cast through the glowing, absinthe leaves.

By the time I was in high school, developers had purchased most of that land. The ridges of the forest were dynamited and pushed into the valleys. The streams where we caught crawdads were forced into pipes that flow underneath parking lots. Where there once was a rich mosaic of woodland plant communities, there are now housing developments and big box retail stores.

My story is not unique. Every day and in every corner of the earth, acre after acre of wilderness disappears. For me, the loss of the only authentic landscape I knew as a child is something that stays with me. It roots me in the reality that the wild spaces we have left are but tiny islands surrounded by an ever-growing ocean of developed landscapes. There is no going back. But the task that faces us now is not to mourn what is lost, but to open our eyes to see the spaces that surround us every day: our yards, roads, office parks, malls, woodlots, parks, and cities.

> Appalachian forest in late winter.

Where heavy industry once charred the ground, wild plants now thrive. The Südgelände in Berlin, Germany, allows us to experience the inexhaustible resource of wildness in the vibrant center of a European metropolis.

CLAUDIA

East Germany in the 1980s was a grey and polluted world. The rivers of my childhood had a different color every week—depending on the color dye used in the textile factories nearby. I remember entire landscapes being ripped out just to get to the shallow layers of soft coal desperately needed to keep a fragile economy afloat, and to pay back reparations for the Second World War. Uranium mines surrounded my hometown and some years (I remember the year of Chernobyl) mushrooms grew twice as large as they normally would. The East German regime relied on intense agriculture practices. Chemical warfare on weeds and pests was so common, nobody ever bothered to bring in the clothes hanging to dry before heavy yellow sprayer planes dumped pesticides across our fields and gardens. Natural areas commonly turned into military training grounds, and nature was reduced to ruderal vegetation and our small but intensely cultivated Schrebergärten.

All this changed in 1989 and 1990 with the fall of the Berlin Wall. The one lesson we learned was how resilient nature really is. Visiting the industrial core of former East Germany now is a life-changing experience: we catch safe-to-eat trout in the once highly toxic streams. Tourists from all over the world come to the new landscape of

The irrepressible spirit of plants: a seeded mix of sweet pea (*Lathyrus odoratus*) and fescues glows in a parking lot in Boulder, Colorado.

Mexican feather grass (*Nassella tenuissima*), catmint (*Nepeta ×faassenii* 'Walker's Low'), meadow sage (*Salvia nemorosa* 'Caradonna'), and alliums intermingle in this mixed perennial bed in author Thomas Rainer's garden.

central Germany—a landscape of clear lakes and shady forests, filled with resorts and expensive yachts. Who would have ever predicted the return of the European wolf to central Europe's new wilderness? Nature is tough, tenacious, and buoyant and it is never too late: this is perhaps the clearest lesson learned when I reflect on the last few years of my young life. Disturbed landscapes heal fast, driven by the powerful and ever-present spirit of the wild. The process of restoration and succession can be very quick if we guide and work with it. Even the most depressing moonscape might be Eden for some plant specialist. Don't tell me you can't find a plant for a challenging site—plants grow on the moon. I have seen it!

. . .

We come to this book representing two different continents and two different experiences of nature. The North American perspective still has a memory of wilderness; the European perspective is immersed in an entirely cultured landscape. Thomas's story is one of nature lost; Claudia's is one of nature regained. This combined perspective perfectly describes the tension in which nature now exists: its continued disappearance in the wild; its expanded potential in urban and suburban areas. Wild spaces may be shrinking, but nature still exists. It is the alligator swimming in the storm water detention pond; it is *Paulownia* growing in the alley; it is endangered sumac that reappears on the military bombing grounds; it is the meadow planted on top of a skyscraper.

We are grounded in the reality of today's environmental challenges. Yet we are entranced by the potential of plants in our human landscapes. And we believe in the power of design. This book is an optimistic call to action, a manifesto dedicated to the idea of a new nature—a hybrid of both the wild and the cultivated—that can flourish in our cities and suburbs, but it needs our help. It requires us to lose the idea that nature exists apart from us, and to embrace the reality that nature in the future will require our design and management.

The front lines of the battle for nature are not in the Amazon rain forest or the Alaskan wilderness; the front lines are our backyards, medians, parking lots, and elementary schools. The ecological warriors of the future won't just be scientists and engineers, but gardeners, horticulturists, land managers, landscape architects, transportation department staff, elementary school teachers, and community association board members. This book is dedicated to anyone who can influence a small patch of land.

< The vast, wild spaces that once covered North America now
exist only in fragments, such as this managed woodland savanna
at the Morton Arboretum in Lisle, Illinois.

INTRODUCTION

NATURE AS IT WAS, NATURE AS IT COULD BE

Imagine for a moment what it must have been like for the first European colonists arriving on the shores of America. The moment they first looked upon the vast, green breast of the continent, their heads full of new world dreams. By all accounts, the landscape they encountered was a place teeming with diversity, a place so resplendent and abundant with life that even our most cherished national parks pale in comparison. Hundreds of species of birds flew over the coastline; tens of thousands of different plants covered the forests, and billions of oysters and clams filled the estuaries. Botanical records and early diaries give us mere glimpses of the richness that once was. Just beyond the coastal plain,

chestnut trees—some nine stories tall—accounted for fully half of the canopy of the Piedmont. These giants showered the ground with their mast, sustaining black bears, deer, turkey, and other creatures. Underneath the chestnuts, rivers of ferns, pools of ladies' slippers and orchids, and sparkling stands of trout lily and false rue anemone— now rare collector's specimens—covered the forest floor. It was a paradise of native species. But to the early colonists, it was a moral and physical wilderness which required great ingenuity and perseverance to tame.

And now we have tamed that landscape. This primal wilderness of our ancestors is utterly gone. Compared with the rich diversity of the past, the modern tableau is a tragedy. Through great engineering and skill, we have drained the Everglades, turned the great American prairie into grids of corn and soybean, and erected Manhattan out of the swamps of the Lenape. The splendor of what once was now exists in isolated fragments, a pale reflection of its former glory.

In this light, the recent rally around native plants bears a bit of irony. The belated rediscovery of the virtues of native plants comes at the moment of their definitive decline in the wild. Conservationists cling to the last slivers of wilderness in nature preserves and parklands. But even these places diminish, as invasive species and climate change alter ecosystems in the most remote corners of the world. To turn back the clock to the landscapes of 1600 is no longer possible. There is no going back.

Of course, there are some success stories of sites being restored to a more so-called native state. But even these successes must be understood in context. Removing invasive

Why not? We may have driven nature out of our cities, but now it is time to think about how we can invite it back. Sumac and little bluestem cover the roof of a gas station.

species can take years of heavy labor or herbicides. Once invasives are removed, sites must be covered with new native plants to keep the invasives at bay. Even then, they rarely stay away. So a site must be continually weeded and replanted, a process that research scientist Peter Del Tredici says "looks an awful lot like gardening." Against the backdrop of species invasion and climate change, these restorations feel quite small—like making little sand castles. All the while, a hurricane gathers on the horizon.

For lovers of nature, this loss creates a deep, collectively shared wound. It fuels a kind of nostalgia for the past, a belief that we can put things back the way they were. In its uglier incarnations, this impulse creates an inflated moralism around the debate over native and exotic plants. What is worse, it makes an ideology out of localism, elevating a plant's geographic origin over its performance.

However, this sense of loss can actually serve a useful purpose. Our mourning creates a craving for an encounter with the natural world. We long to feel small in the midst of an expansive meadow, to witness the miracle of a moth emerging from a cocoon, or to be filled with the glow of morning light on beech leaves. Our ancestors experienced these events as a regular part of their days, but now our children often learn these moments only through YouTube. We hunger for an authentic connection with the landscape that engages our senses and fills us with wonder.

A NEW OPTIMISM: THE FUTURE OF PLANTING DESIGN

A new way of thinking is emerging. It does not seek nature in remote mountain tops, but finds it instead in the midst of our cities and suburbs. It looks at our degraded built landscapes with unjaded eyes, seeing the archipelago of leftover land—suburban yards, utility easements, parking lots, road right of ways, and municipal drainage channels—not as useless remnants, but as territories of vast potential. We pass them every day; their ordinariness is what makes them special. As such, they are embedded in the fabric of lives, shaping our most recurring image of nature. French landscape architect Gilles Clément calls these fragments the Third Landscape, the sum of all the human-disturbed land through which natural processes still occur.

For designers, the loss of nature is a starting point. It helps us to look at our cities with fresh eyes, giving us a sort of x-ray vision that cuts through the layers of concrete and asphalt to see new hybrids—of natural and man-made, of horticulture and ecology, of plant roots and computer chips. It allows us to imagine meadows growing on skyscrapers, elevated roads covered with connected forests, and vast constructed wetlands that purify our drinking water. But this future will not be driven by the assumption that what is natural is only that which is separate from human activity. Instead, it begins with the conviction that all naturalism is really humanism. Only when we clear our heads of the rose-tinted idealism of the past can we really embrace the full potential of the future.

To get to that future requires serious work, serious engineering, and serious science. But it does not require our plantings to be so serious. In an era of climate change and species invasions, the only certainty is a whole lot more uncertainty. The high-maintenance lawns and clipped shrubbery of office parks and suburban yards will seem increasingly odd with every large-scale natural disaster or water shortage. Since we will not have absolute control, planting in the future will become more playful. More whimsical. Faced with a landscape of increasing instability, planting no longer has to be so solemn. It can loosen up. Be more frivolous. The uncertainty of the future will provide an incredible gift: it will liberate planting from all those forces that try to tame it—the real estate industry, "good taste," designers' egos, eco-evangelism, and the horticultural industry. It frees us to take risks, act foolishly, and embrace failure. After all, no designed planting ever lasts. Its main purpose is not to endure but to enchant.

So what exactly is the planting of the future? Look no farther than just outside your front door. Go find a patch of weeds in your neighborhood. Notice the variety of

Mayapple (*Podophyllum peltatum*) pools beneath the canopy of an emerging oak tree, conveying a fittingness of plant to place.

species and how they interweave to form a dense carpet. Or better yet, take a hike in a nearby natural area. Look closely at how plants grow in a meadow or a forest's edge. Observe the lack of bare soil and the variety of ways plants adapt to their site. Then when you get back to your neighborhood, compare those wild communities to the plantings in landscape or garden beds. There is a difference between the way plants grow in the wild and the way they grow in our gardens. Understanding this difference is the key to transforming your planting.

The good news is that it is entirely possible to design plantings that look and function more like they do in the wild: more robust, more diverse, and more visually harmonious, with less maintenance. The solution lies in understanding plantings as communities of compatible species that cover the ground in interlocking layers.

v A patch of weeds densely colonizes a narrow sidewalk hell strip. Over twenty species of plants—mostly exotic—thrive in this small space.

> The repetition of cinnamon fern (*Osmundastrum cinnamomeum*) gives structure and interest to this cranberry glade. Underneath, grass pink orchid (*Calopogon pulchellus*) mixes with wild cranberries (*Vaccinium* sp.) and sedges.

BRIDGING THE GAP BETWEEN NATURE AND OUR GARDENS

The way plants grow in the wild and the way they grow in our gardens is starkly different. In nature, plants thrive even in inhospitable environments; in our gardens, plants often lack the vigor of their wild counterparts, even when we lavish them with rich soils and frequent water. In nature, plants richly cover the ground; in too many of our gardens, plants are placed far apart and mulched heavily to keep out weeds. In nature, plants have an order and visual harmony resulting from their adaptation to a site; our gardens are often arbitrary assortments from various habitats, related only by our personal preferences.

For too long, planting design has treated plants as individual objects placed in the garden for decoration. Unrelated plants are arranged in ways that are intended to appear coherent and beautiful. To assist designers and gardeners with this difficult task, there are endless books on plant combinations, perennial borders, and color harmonies. The heaving bookshelf of garden books leaves us with endless tips and information, but very little real understanding of the dynamic way plants grow together.

Not surprisingly, this individualistic approach is also high maintenance. Each plant has different needs: some need staking, others need more water, yet others need soil additives. In fact, the very activities that define gardening—weeding, watering,

fertilizing, and mulching—all imply a dependency of plants on the gardener for survival. Gardeners are often frustrated when some plants spread beyond their predetermined location and surprised when others struggle to get established. Many come to believe that successful gardening is only for those with a magical touch, a green thumb, or some other mystical insight bestowed upon the chosen few.

A further complication is the availability of plants from every corner of the globe. Plant selection is often overwhelming, providing infinite choice but little real sense of how to create stable and harmonious planting.

THE INSPIRATION OF NATURALLY OCCURRING PLANT COMMUNITIES

So how do we shift the paradigm, making desirable plantings that look and function sympathetically with how they evolved in nature? By observing and embracing the wisdom of natural plant communities.

Wild communities differ from our gardens. They are better adapted to their sites, more richly layered, and have a strong sense of harmony and place. For designers, these qualities are highly desirable. But to achieve them, we must arrange plants to interact with other plants and the site, understanding the wide variety of roles plants play in a community. Some cover the ground in large colonies; others exist as solitary specimens. Some take up excess nutrients; others add nitrogen to the soil. Through years of competition and natural selection, plants segregate themselves into these different roles to make the best of limited sun, water, and nutrients. These communities are functional workhorses, performing valuable ecological services that far surpass conventional plantings. The end result is a rich mosaic of species, exquisitely tuned to a particular site.

It is natural to look upon wild plantings with a kind of supernatural reverence; even ecologists admit how little we know about plant interaction within communities. But our sense of wonder about these dynamics need not blind us to the lessons they offer. We can look at wild plant communities and see a few principles at work that can help designers better select, arrange, and manage horticultural plantings.

This book is a guide for designing resilient plant communities. We want to demystify the process of creating stylized versions of naturally occurring plant communities that work in tough, populated sites, offering inspiration for arranging species in ways that work with a plant's natural tendencies—its evolved competitive strategies—rather than fighting them. Here you will find the tools to select the right plants for a site, to vertically layer plants in a composition, and to translate naturalistic planting into visually compelling compositions. For those interested in real-world ways to accomplish lusher planting with less input, this book offers a simplified, practical method that seeks to both please humans and sustain fauna.

Designing with plant communities can not only link nature to our landscapes, but also bring together ecological planting and traditional horticulture. In the last decade in particular, a troublesome divide has developed between those interested in native and ecological planting and those immersed in traditional gardens and horticulture.

Symphyotrichum, Solidago, and *Pycnanthemum muticum* form an ecologically valuable alliance in this natural meadow community.

Adam Woodruff's design for an Illinois residence artfully mixes U.S. native genera like *Silphium, Baptisia, Echinacea,* and *Sporobolus* with exotic *Perovskia* and *Calamagrostis* ×*acutiflora* 'Karl Foerster'.

The debate over the use of native and exotic plants in particular has polarized gardeners. It makes some feel judged for not being "green" enough, and others persecuted for caring about the environment. What could be an important dialogue is too often reduced to inflated ideology. Worst of all, the debate is so focused on *what* to plant that it almost never addresses the more important question for gardeners and designers of *how* to plant.

The idea of designed plant communities offers a middle way. It provides real solutions for the central concern of native plant advocates, providing more diversity and better ecological function. The focus on layered plantings means that there can be more beneficial plants in small spaces. Yet it also acknowledges our contemporary dilemma of wanting to create more "nature" in landscapes that no longer resemble historically natural conditions.

This middle way looks with fresh eyes on two very different types of plant communities. One type is the native plant community, such as the historic ecosystems of our last remaining wilderness areas. Thousands of years of competition and evolution have produced these environments with a remarkable degree of beauty, harmony, and order. The other type of inspiration includes naturally occurring cosmopolitan communities, such as common weed patches. Take a walk down your street and you will likely encounter numerous examples of plants that have spontaneously colonized the derelict cracks and corners of the neighborhood. Even in the harshest urban conditions, plants abound in roadside medians, empty woodlots, margins of parking lots, the gravel beds of railroad tracks, and in compacted lawns.

Our cultural associations with these two categories of plants could not be more divergent. We celebrate the native plants of our forests, meadows, and deserts. Millions of dollars a year are spent studying and protecting these fragments of wilderness. At the same time, we vilify the weeds that emerge in our "civilized" landscapes. We spend endless hours pulling, spraying, and mulching to prevent them. What we see in each of these plant communities—both the beloved and the despised—is the way groups of related plants adapt to their sites. Both examples are models of how a diverse mix of species inhabits different niches. And both are incredibly tough, resilient, and self-sustaining. This is not to say that both are equal in terms of ecology or even beauty. In fact, our cultural preference for one over the other is indeed meaningful, and should influence design.

By focusing on *naturally occurring* plant communities, as opposed to those that are *purely native*, the focus is shifted from a plant's country of origin to its performance and adaptability. This shift is absolutely crucial. At the same time, our willingness to consider more than just native plants does not mean they have no place in the futures scheme. We firmly believe that designing with native plants still matters. In fact, it matters more than ever. But in order to be successful in establishing native communities in tough sites, both a new expression of nature and a deeper understanding of the dynamics of plant communities are required. It is our challenge to reimagine a new expression of nature—one that survives within our built landscapes, and at the same time performs vital ecosystem functions needed to ensure life. We must put aside our romantic notions of pristine wilderness and embrace a new nature that is largely designed and managed by us. The building blocks of this new nature are resilient and native plants—and yes, even exotic species—that are naturally adapted to environments similar to our man-made landscapes. The question is not what grew there in the past but what will grow there in the future.

For us, the most compelling reason to consider designed plant communities is not ecological or functional, although those are valid and powerful reasons. The more persuasive argument is aesthetic and emotional.

Most of us live in landscapes created and managed by people. In contrast to the spirited spontaneity of wild vegetation, the landscapes of our yards, office parks, and cities are plastic assemblies of overused evergreens sheared into meatballs and vast seas of mown lawns. If any color is used at all, it is with flimsy rows of bedding annuals. There is a harnessed quality to these plantings that results in limp, hollow-feeling landscapes.

CONNECTING WITH OUR MEMORY OF NATURE

The nature we do experience is limited to our small parks and yards, often overlooked and not large enough to offset the negative effects of the built environment. Yet we are deeply connected with nature, and remember a past when nature surrounded us and played a larger role in our lives. We no longer sleep under the stars, break the soil with our hands, or read the plants in the forest to find our way home. But a part of us still longs for that connection. It is only in the last hundred years or so of our species that we have become removed from our outdoor environments. It is not that we have lost the capacity to read and see landscapes, but we are out of practice and we are desperate for it.

At a deep level, when we see plants that perfectly fit their environment, it reminds us of an ancient fellowship we had. The immense popularity of the High Line park in Manhattan—now one of the most popular landmarks in the world—underscores the desire to experience places within our built world that remind us of wildness. We go on extensive hikes in parks or enjoy the wilderness of alpine regions from our mountain bikes. The natural landscapes we seek seem to have an emotional pull on us. They make us breathe deeper and balance our spirits.

Capturing the spirit of wild meadows atop Pittsburgh's Convention Center. *Liatris spicata* mingling with *Penstemon digitalis* and grasses.

There are deep evolutionary reasons why natural forms resonate so strongly and why certain plant communities are commonly perceived as harmonious and beautiful. Their ranges of colors, textures, and seasonal expressions please our eyes and can have therapeutic effects on us. The work of environmental psychologists suggests that our attraction to specific modern landscapes such as parks with trees and lawn may arise because they evoke environments that supported and fed our ancestors thousands of years ago. By contrast, many of our built spaces lack those profound triggers and consequently the emotional responses. In constructed settings, we have so few places in which we experience beauty, and as a result, we rarely develop a deep and meaningful relationship with them.

This is the missed opportunity. Truly great planting reminds us of a larger moment in nature—when a group of garden plants makes you feel like walking through a meadow, or hiking through a dark forest, or entering into a woodland glade.

Here we propose a method for creating designed plant communities that work in both urban and suburban sites. The first chapter explains the idea of designed plant communities and introduces its essential principles. The second chapter looks at the inspiration of nature and grounds the reader in the dynamics of plant communities in the wild. In the third chapter, we describe the design process: how to understand your site, develop a palette of plants, and arrange and layer plants. Finally, the fourth chapter examines the unique installation and management requirements of designed plant communities.

More than ever, we need planting solutions that are resilient, ecologically functional, and beautiful. Our goal with this method is not merely to create more functional plantings, but to make people see again, to make them remember. We arrange plants in ways that will conjure experiences of the ephemeral. It is not the plants themselves that have power; it is their patterns, textures, and colors—particularly those that suggest wildness—that become animated as light and life pass through them.

∨ A rooftop meadow in Manhattan brings the warm colors of native grasses and forbs to the skyscrapers of New York.

∨ Experiencing the unfettered essence of the prairie, high over busy streets.

Lowland wild flowers and moist summer hay meadows inspired Sarah Price and Nigel Dunnett's designed plant community for the Olympic Park European Garden (part of what is now Queen Elizabeth II Olympic Park in London).

PRINCIPLES OF DESIGNED PLANT COMMUNITIES

Loosening the grip on our cherished notions of plant arrangement makes it possible to transform our adversarial relationship with nature into a collaborative one. The dry, gravelly hell strip along the edge of a street can be busted up, enriched with compost, and planted with boxwood and hostas. Or it could be preserved exactly as it is and made into the perfect home for drought-loving plants like Mediterranean mints, low meadow grasses, desert annuals, spreading sedums, and alliums. Intimately understanding a site is part of the challenge, but the more important task is to understand how plants fit together.

DEVELOPING AN UNDERSTANDING

Plant communities are human constructs, conceptual frames for describing a group of plants in a place. The term is an ecological one, but increasingly designers have adopted it to describe composed plantings. Over the last century, the concept of a plant community has evolved significantly as scientists gained new insights into the way plants interact with each other and their sites. So it is important to ground our understanding in current ecological principles, but still work with the term to make it useful and relevant for design.

These communities do not exist in nature as distinct organisms, as many ecologists once believed. Earlier twentieth-century theorists tended to idealize plant communities as kinds of superorganisms, believing that species cooperated for the benefit of the group, like ants in a colony or bees in a hive. Early concepts also thought that plant communities had distinct boundaries, with zones of transition (ecotones) between them. However, the modern understanding of plant communities is very different. Today, there is wide agreement among ecologists that the composition and boundaries of most communities are fluid. This is because each species responds individualistically to its site. While certain biological interactions are indeed mutually beneficial (i.e., mycorrhiza to plant roots), research does not support the idea that a community is some kind of tightly linked superorganism, but is instead made up of groups of overlapping populations that coexist and interact.

Plant communities are an abstraction, a naming convention we use to describe vegetation so we can study it. These conventions are built using various classification systems, with each system describing different characteristics. Some focus on geographic and climatic boundaries; others on dominant plants. They can vary in scale from large biomes to very specific vegetation patterns. Classification systems have their strengths and weaknesses, and can be used in combination with each other if needed. Because they are our own construction, there is no right or wrong way of classifying plant communities. Our use of the term plant community focuses on smaller scale groups, mostly because these are more relevant to designers. Plant communities can be classified using various systems and scales, ranging from broad biome classifications to detailed analyses of local plant communities.

Within the conceptual frame of a plant community, the plants are merely a snapshot of those growing together at a certain point in time. Over the procession of seasons, plants associate and disassociate freely with one another. Many of the plants we commonly see together in the wild today likely were not found together before glaciation or even as little as a century ago. For example, white pines, hemlock, chestnut, and maple are common associations in the last 500 years, but before glaciers pushed through the upper midwestern part of the United States, they would not have been found together. Moreover, new plant communities appear every day. The introduction of exotic species leads to previously unseen plant combinations. Plants that never saw each other before often thrive in what we call novel plant communities. Think of milkweed mingling with European cool season grasses in rural pastures of the eastern United States—a rather common site today.

> Colonies of bracken fern (*Pteridium aquilinum*), gold in autumn, work as a seasonal theme layer in this community, while species of *Carex* and *Vaccinium* form a ground-holding matrix.

Plant communities vary greatly in how quickly they change. Their stability has a lot to do with their surrounding climate and the stability of their habitat. Some fire ecosystems transform from one vegetation pattern to another in just a few months. Far-north plant communities with short growing windows, on the other hand, can be surprisingly stable. Slow transformations caused by shifting climate and succession are hard to observe. But we certainly notice the more obvious short-term changes.

There is no limit to the number of successful plant communities possible on earth. What we see growing together now is only one possible version; countless other plants will work together if they have the chance to meet in the wild or in cultivation. Populations can only develop on a site if their seeds or parts of their roots were blown or carried there by natural forces, animals, or humans.

The fact that we see plant populations growing together has as much to do with chance as with adaptability. Populations are distributed along gradients of environmental

∨ Coastal woodlands are prone to fire. Every time they burn, the herbaceous ground layer changes. Some species are encouraged by fire and rejuvenate quickly; others disappear forever.

∨ Stable plant community in a rock formation. This community likely looked almost exactly the same decades ago.

Stable Climax or Constant Flux?

Past theories described the way plant communities age in discrete stages of succession, ultimately resulting in a predictable and stable "climax" community. A general consensus has emerged over the last few decades that most communities never reach a climax or stable equilibrium; instead, they are constantly changing as a result of disturbances such as plant death, fire, wind, ice, water, and even human activity. When a tree falls in a forest, or an invasive plant displaces a native one, or a road is cut through wildlands, the process of succession starts again. Change is constant.

PLANT POPULATION CURVE

Plants respond individualistically to a site because species have different tolerances to site constraints such as shade, drought, or soil infertility. They flourish in their optimum ranges, and numbers decline moving away from ideal conditions.

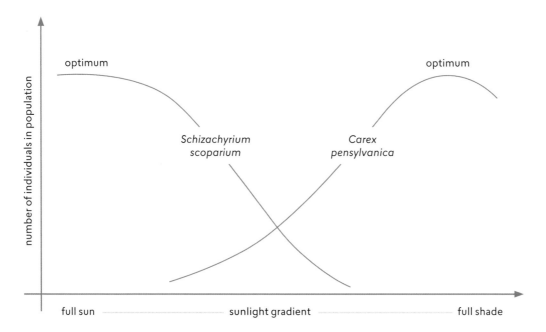

COEXISTING POPULATIONS

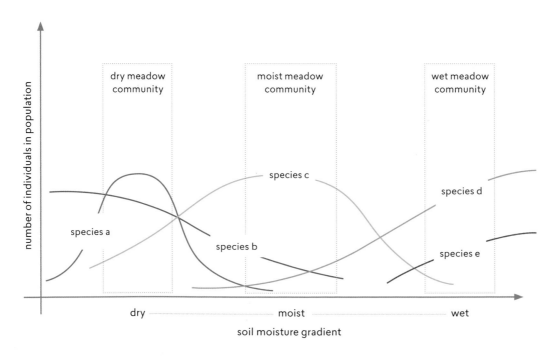

PLANT COMMUNITY CLASSIFICATION SYSTEMS

CLASSIFICATION SYSTEM	CHARACTERISTICS	EXAMPLES
Physiognomy	Large scale biomes based on plant physiognomy. Describes vegetation worldwide.	Tropical rain forest, temperate forest, taiga
Dominant species in tallest layer	Dominant species in tallest layer used to differentiate communities. Describes regionally occurring communities.	Oak-hickory forest, red maple woodland
Dominant species of each layer	Small-scale description, dominant species in each layer is captured in plant community name. Describes very specific local plant communities.	Chestnut oak/lowbush blueberry/hay-scented fern forest

ROOT MORPHOLOGIES

Plant species have different root morphologies, allowing them to access water and nutrients from different soil horizons. Each root system occupies a different belowground niche, limiting competition between species. For example, deep taproots of tall forbs do not directly compete with the shallower, fibrous root systems of short grasses and forbs.

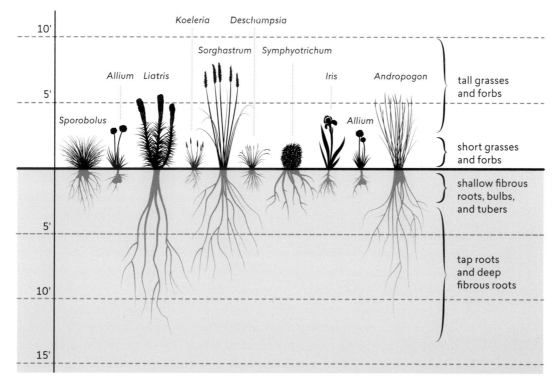

Ironweed (*Vernonia noveboracensis*) and common boneset (*Eupatorium perfoliatum*) aggregate along the wetter portions of this pasture.

conditions, such as soil moisture from wet to dry, or topographic elevation from deep valleys to high mountaintops. Plant populations do not grow equally well along the entire gradient. For example, moisture-tolerant perennials will fade out as one moves to gradually drier portions of a site. A shade-tolerant plant will become more prevalent the deeper one goes into a forest. Species thrive in what we call their optimum within a gradient, and they struggle to hang on toward the extreme ends of their distribution area, because of increasingly unfavorable growing conditions. For example, *Carex stricta*, *Juncus effusus*, and *Vernonia noveboracensis* thrive in the moist center of a wetland but struggle to survive in the surrounding drier areas.

Beyond adapting, plants have to be able to compete with other populations in order to survive. Resources such as light, water, and nutrients are limited and plants fight for their share in order to survive and reproduce. Young plants only successfully establish if they can compete with other populations growing on the same site. Not every population is compatible with another. Plants coexist because they occupy various ecological niches within their environment. These specialized niches allow different plants to make use of limited resources in what seems to be exactly the same spot. Plants exploit different spaces with a number of remarkable alterations, such as varying rooting depth, height, moisture tolerance, light tolerance, or by teaming up with microbes to help them get nitrogen from air instead of soil. If plants occupy the same niche, they compete directly with one another. Examples of direct competition are plentiful in

Lush or sparse

> Plant communities can be lush or very sparse in vegetation. Think of a thick carpet of herbaceous species in a mesic meadow.

>> Compare that to the rather sparse vegetation of serpentine barrens—a community with lots of open ground due to extreme site conditions and frequent disturbance.

Few species or many

> Plant communities vary in levels of diversity. A saltwater marsh can be dominated by only a few species, such as common reed and cord grass.

>> Conversely, a meadow community can have dozens of species in one square meter. Most vegetation patterns are formed by overlapping species populations.

Varied morphological expressions

> Plant communities have countless morphological expressions. The spiky leaves of *Yucca filamentosa* create one.

>> Another expression is created by soft fern fronds.

planting. *Panicum virgatum* is a grass that can easily reach six feet high in the wild, but when crammed into a dense, monocultural planting, it may only reach four feet tall and be covered in stress-related leaf rust disease. Direct competition can cause stunted growth as well as poor plant development and health.

The modern understanding of plant communities reveals a complex web of interwoven relationships between plants and place. Even modern ecology has yet to fully understand all the intricacies of plant interaction. However, we do not need complete comprehension in order to create plantings that function more naturally. What is important is to understand the essential elements that define a plant community, and to use those elements to create more resilient planting.

WHAT MAKES A DESIGNED PLANT COMMUNITY DIFFERENT?

A designed plant community is a translation of a wild plant community into a cultural language. Why do plant communities need translating? Practicality, for one thing— urban and suburban landscapes are so drastically altered from the historic ecosystems that once existed. Think of your home and then think of the landscape that existed there a thousand years earlier. The process of urbanization has entirely altered the environmental conditions. So a designed plant community may reflect these changes by incorporating a narrower selection of the most adaptive species. Or it may include species from different habitats to supplement a native palette, particularly when an all-native selection is not commercially available.

The second reason we create designed interpretations is to increase the pleasure and meaning of the plantings for people. This could involve increasing the number of flowering species to make the community more colorful. Or simplifying the palette of plants and exaggerating the natural patterns to make the plantings more ordered and legible. A grassland-inspired design may place accent perennials tighter together to make drifts even more noticeable. Or a single species of understory tree might be repeated in a woodland planting to create a more dramatic effect in spring. Amplifying the signature patterns of a plant community helps to make them more readable and enjoyable.

Designed plant communities represent a hybrid of horticulture and ecology. Because of this, we want to distinguish the creation of designed plant communities from ecological restoration. While designed communities may indeed provide many ecological services, they are not necessarily true ecosystems. We are optimistic about the ecological potential of designed communities, but some humility is still needed. Naturally occurring plant communities are the result of millions of years of natural selection and succession. It is doubtful that any designed planting plan could replicate all the dynamics of a real ecosystem. We still have much to learn. So until the research is more advanced, we consider designed plant communities to still reside more in the realm of horticulture than ecology.

∨ Tight drifts of *Persicaria bistorta* and *Iris sibirica* seem to float in a matrix of *Molinia caerulea* at the Trentham Estate, a stylized representation of the patterns of a meadow.

⋁ The show garden of Lianne Pot in the Netherlands features prairie-inspired perennial meadows. The modular approach repeats a mix of dominant theme plants with companion plants. Unlike their wild counterparts, they do not evolve, but are assembled and tended by people.

⩘ The harmony of colors and textures, down to the smallest detail, is a unique byproduct of using native plants that have evolved under similar conditions.

⋀ Mosses, lichens, grasses, and tree seedlings colonize the cracks of a boulder in Shenandoah National Park.

⩘ Appalachian rockfern (*Polypodium appalachianum*), a recently recognized species, grows in the rocky slopes of its namesake mountains.

⋀ The North American palette of native plants is large and includes colorful choices such as *Rudbeckia* and *Heuchera villosa* var. *villosa*, shown in Sarah Price and James Hitchmough's design for the Olympic Gardens North America.

The concept of a designed plant community is entirely agnostic about where the plants come from. The group can be composed of an international mix of species or an entirely native palette. In fact, a plant community may be composed of all exotic species and still engage in ecological processes similar to a naturally occurring community. This viewpoint differs from a small but vocal faction of the native plant movement that characterizes all non-native species outside the realm of ecology. This is simply not true. *All* species—native and exotic—have specific ecological niches and interact with their environments and other plants. The notion of the innate superiority of native plants is problematic in that it ignores the reality that our towns and cities are increasingly surrounded by non-native vegetation. While there may indeed be ecological benefits specific to certain native species, exotics can play important roles in the formation of plant communities. The obvious exception is when the plant possesses the potential to spread beyond the site and displace or disrupt local native plant communities.

THE ROLE OF NATIVE SPECIES

Our focus is on plants naturally adapted to their specific sites. It is the *relationship of plant to place* we want to elevate. For this precise reason, native species can and perhaps should be the starting point for developing high-quality designed communities. In many ways, starting with a native plant community as a reference point can simplify the design process.

In order for a designed planting to become a community, two conditions must be met. First, all plants chosen should be able to survive in similar environmental conditions. A desert agave and wetland iris, for example, obviously would not work together to form a self-sustaining community. Compatible species should be able to grow and thrive within the same environmental stresses and disturbance regimes. The second precondition of a plant community is that the plants must be compatible in terms of their competitive strategies. Understanding these different competitive strategies is the key to plantings that last.

Native plant communities offer an inherent advantage with regard to both these conditions. Simply put, plants that grow together in the wild will likely go together in a similar landscape setting. While it is indeed possible to substitute an exotic plant that may also be adapted to the same conditions, choosing plants beyond the range of an existing plant community increases the burden on the designer to understand how the plant will perform in a novel community. Combinations that already exist in nature are somewhat battle-tested. Many such associations have endured for thousands of years. The more our combinations differ from natural combinations, the greater the risk.

Perhaps the most compelling reason for starting with native plant communities is to give the site a sense of authenticity. The long, organic adaptation of plant to place produces a harmonic relationship that is almost impossible for a designer to fully replicate. Consider the beauty of even the smallest moments in the wild: how the bright colors of lichens and mosses balance the more neutral colors of rock outcrops and dried grasses that surround it; how the contrasting textures of ferns against a craggy shrub heath create a playful rhythm in a wet meadow; and how the silhouettes of dried seed

heads pierce through the misty inflorescences of grasses. It is the accumulation of all these details that conveys a spirit of place. Of course, a well-designed exotic ensemble can also tickle our memory of nature, but this almost entirely depends on the skill of the designer. Starting with a palette of plants that naturally evolved together just simplifies the task. Native plant communities often have all the various components—a complete ingredient list—necessary to execute a recipe for resilient and stable design.

Designed plant communities place the emphasis on a plant's ecological performance, not its country of origin. We are interested in practical solutions, not ideological dogma. The combination of adapted exotics and regionally native species can expand the designer's options and even expand ecological function. This gives the designer great flexibility to blend a variety of species to create likenesses of natural plant communities, including combinations that may not actually grow together in nature.

< Plants have evolved to grow among other plants, not as lone specimens. *Typha latifolia*, several species of *Scirpus* and *Carex*, and *Eupatorium perfoliatum* mingle on the edge of this pond.

Our simplified, practical methodology focuses on five key principles that define the essence of a designed plant community.

These principles are applicable to any designed planting, regardless of style. Stylistic preferences for gardens vary along a gradient from formal to naturalistic. Our purpose is not to endorse any one particular style; a designed plant community may be highly naturalistic, but it might also be formal or modern. Any kind of design can benefit from combining plants more as they exist in nature. What matters is that plants are allowed to respond to the site and have some role in shaping their own destiny.

ESSENTIAL PRINCIPLES

PRINCIPLE 1: RELATED POPULATIONS, NOT ISOLATED INDIVIDUALS

Moving from the idea of a traditional planting to a designed plant community starts with letting go of the idea of plants as objects to be placed, like pieces of furniture. Instead, think about plants as groups of compatible species that interact with each other and the site. To understand this distinction, consider for a moment the tale of two plants, a wild species and a cultivated one.

A wild plant is self-planted. It was grown either by seed dispersed from a nearby plant or vegetatively through an adjacent plant. For a young plant, the road to maturity is treacherous. Many perish. Some die when more established plants smother them for light; others die for lack of water or nutrients. The plants that do survive must find a space that no other plants inhabit. Faced with death, they adapt. They creep through the gaps between other plants; they delay their growth for the most opportune time (bulbs, cool and warm season grasses); they change shape to fit between other plants, or they form root structures that allow them to compete with larger plants. This process is slow and iterative. Yet over time, the end result is an extravagance of planterly life, intricately woven through and around each other.

In contrast, a cultivated plant is propagated in a nursery that artificially controls light, nutrients, and temperature. The nursery places the plant in a peat-based soil mixture and waters and fertilizes it until it is ready to be sold. The plant is then chosen by a gardener who takes the plant and installs it in amended soil, often without understanding how the soil's pH or fertility will affect the plant. Placement is most often determined by where we think they will look good, often for some ornamental conceit, like a color theme. Plants are generally placed far apart to avoid competition, and mulched heavily to prevent weeds. Unless the gardener is highly knowledgeable about a plant's cultural requirements, the result is typically a random exhibit of plants from different habitats.

Traditional horticulture arranges plants as isolated individuals. Paired with the right companions, the best qualities of this *Acanthus mollis* could be amplified.

A painterly approach to plant composition arranges plants like paint chips. All evocative connections between individual plants and between plants and their natural environment are lost. The sharply contrasting foliage here creates a jarring effect.

These two tales reveal the sharp differences in the ways groups of plants react to a site. The first is a story of plants shaping their own destiny by conforming to a specific site. The remarkable story of dispersal, establishment, competition, and adaptation highlights the relationship of plant to place. The second story brings to light the rather arbitrary way humans arrange plants in our gardens. In this story, plants have little, if any, control over their own fate.

Matching groups of plants to their site is a seemingly common sense approach. Yet the extent to which this simple precept is ignored by professional designers is alarming. The haphazard way most plants are matched to a site is entirely embedded in our landscape culture. Planting design is often taught in the language of abstract formalistic composition, devoid of basic principles of vegetative ecology. Analogies of planting to painting abound, implying a two-dimensional relationship of plant (paint) to site (canvas). Planting curricula at most landscape architectural schools in America focus on a handful of overused woody plants in a "Plant Materials" course—a term that reveals the rather static view in which plants are regarded. These courses tend to emphasize ornamental characteristics, but almost no information is given about how plants mix together, what kind of roots they have, or how they compete. Beyond design pedagogy, entire libraries of garden books teach individualistic plant arrangement; amending soil to suit a heterogeneous mix of plants; and a reliance on mulch, irrigation, and fertilizers to keep plants alive. The irony is that in focusing on the ornamental features of plants exclusive of their ecological traits, we're not developing the skills in combining plants that are necessary to achieve the design effects we want.

To liberate your plantings, consider a plant as a piece of a very large puzzle. In fact,

A weedy spot of lawn is a great place to study the interwoven nature of plants in a community. Here a sea of buttercups (*Ranunculus repens*) creates a seasonal theme in spring.

the analogy of a puzzle is a simple way of describing how a plant's shape—its morphology—is a response to its environmental conditions and other plants.

Consider for a moment a patch of weeds. In an effort to take advantage of bare soil, early colonizers display a spectacular diversity of shapes and textures. The ferny texture of yarrow, the broad basal foliage of greater plantain, the upright blades of bluegrass, and the thick mat of ground ivy all combine in the interlocking puzzle, allowing a remarkable density of plants to coexist. It is not just the plants' foliage that interlocks, but also their roots. Many species are known to have a high degree of root plasticity, the ability to move into different parts of the soil to avoid competition with other roots. The above- and belowground layering of diverse foliage and root shapes makes the puzzle successful.

Plant life in nature is by definition communal. Sometimes plants depend upon other plants (parasites, epiphytes, climbing plants); other times those unions are symbiotic (mycorrhizal fungi and tree roots); yet other times those unions are competitive. In traditional planting, plants are often spaced far apart to avoid interaction, giving gardeners control over the outcome. However, if the right choices are grouped together,

designers can work with each plant's competitive strategies to produce greater effects: a longer succession of bloom, more diversity of texture, and longer-lasting ground cover. Designed plant communities work on the principle that more can indeed be more. When plants are paired with compatible species, the aesthetic and functional benefits are multiplied, and plants are overall healthier.

PRINCIPLE 2: STRESS AS AN ASSET

The curse of temperate climates with rich soil is that one can grow anything. For designers interested in creating communities with a rich sense of place, the first step is simple: accept the environmental constraints of a site. Do not go to great effort and cost to make soil richer, eliminate shade, or provide irrigation. Instead, embrace a more limited palette of plants that will tolerate and thrive in these conditions.

Plants in the wild are inextricably bound to their environments. Think about how the smallest fold of topography in a meadow creates a drift of one species, or how a fallen tree in the forest allows a pool of light from which new species emerge. Plants and the patterns they create articulate even the most subtle changes in land.

Each site, with its unique soil conditions and light levels, favors plants with specific shapes and functions. The apparent harmony that we perceive between plants and their environment is a result of a rather brutal process of natural selection. Each population of plants produces more offspring than can possibly survive. Only the fittest live, resulting in new plants more adaptive than their parents to the local ecological niche. Eons of natural selection result in plants with remarkable site-specific features. Prairie grasses can have roots that are more than ten feet deep, allowing them to regenerate after fire. Some dune species have long taproots and floating seeds that allow them to colonize barren, sandy berms. In dry climates, hairs on some leaves trap moisture from humidity and form a boundary layer to protect the plant from drought. Everything about a plant—its shape, root system, leaves, and reproductive strategy—is a response to a particular site.

From a design perspective, what is so desirable about naturally occurring plant communities is a plant's fitness to a specific site. We admire the way trilliums pool between the roots of an oak, and how coneflowers drift through a meadow. In these communities, there is a sense of spontaneity and harmony that is the result of a plant establishing in a site that can support it. The irony is that what we perceive as happy, well-adjusted plants is more often the result of a scarcity of resources rather than an abundance of it.

A plant's placement on a particular site is a result of its tolerance to the environmental conditions of that site. Tolerance is a key concept here because it describes the plant's accommodation of a limited resource. All vascular plants require basic resources to live: nutrients, water, light, and carbon dioxide. The supply of any of these elements is greatly affected by temperature, pH, humidity levels, and the aerobic condition of the

< Echinacea simulata blooms in an atypical calcium-rich prairie. This soil would be problematic by most horticultural standards, yet its unique qualities support over forty rare and endangered animals and plants.

∨ The site qualities we work so hard to fight are the very qualities that can make for a resplendent planting. Here, brutal drought and infertility help create a plant palette that enhances the beauty of the Arizona desert.

∨ The harmony of vertical shapes and colors of saw palmetto (*Serona repens*), blue broomsedge (*Andropogon Virginicus* var. *glaucus*), and longleaf pine (*Pinus palustris*) is a direct adaptation to the infertile, sandy soil; drought; and brackish water of this Gulf Coast swale.

soil. Unlike animals that can change location to seek food or water, a plant is sessile—it cannot move. When a plant is separated from the resource it needs, it must develop adjustments in its shape, photosynthetic metabolism, or nutrient uptake in order to survive. So if a plant on the forest floor needs to capture more light, it must allocate more of its resources to developing stems and leaves, or add more chlorophyll to its cells. When a plant adjusts itself to seek a limited resource, it does so at the expense of other resources it can acquire. There is an unavoidable trade-off that is a result of a plant's separation from resources.

So it is not just the availability of resources on a site that determines plant allocation, but the lack of it. In a sense, each site—with its unique light levels and soil resources—predetermines the plants that will grow there. The site favors plants with certain shapes and photosynthetic adaptions. A plant's tolerance to different kinds of stress—such as low light, water, or nutrients—will to a large degree influence its distribution on a site.

The takeaway for designers is simple: stress is an asset. Our initial instinct in preparing a site is often to eliminate the constraints we think will limit plant growth. We bust up soil and add organic matter; we remove shade to let more light in; and we install irrigation to provide plants constant moisture. But in many ways, we are obliterating the very qualities of a site that will create a strong sense of place. Traditional garden lore teaches that any soil that is not rich, black loam needs to be improved. Tell that to the wildflowers that thrive in some of the world's most inhospitable soils. In amended soils they often die within just a few years. It is no coincidence that gardens that have the strongest sense of place often have sites with extreme constraints. Englishwoman Beth Chatto's iconic Gravel Garden is celebrated around the world for its wonderful sense of place. It is a garden with poor, gravelly soil that has never been artificially watered. Her garden combines plants from beachside dunes, alpine rockeries, Mediterranean cliffs, and dry meadows to create a lasting community rich in evocative appeal.

Thick, interwoven layering of compatible plants is the hallmark of a plant community—density that is all too uncommon in traditional plantings. Phlox, geranium, trillium, dandelions, and grasses form a close-knit cluster at the base of a tree.

Green mulch. As light levels drop under a tree, grasses transition to a mass of ferns, maintaining a continuous sea of plants.

PRINCIPLE 3: COVER THE GROUND DENSELY BY VERTICALLY LAYERING PLANTS

The approach to ground cover is, for us, the single most important concept of creating a functioning plant community. Think about seeing plants in the wild: there is almost never bare soil. With the exceptions of deserts or other extreme environments, bare soil is a temporary condition. Yet in our gardens and landscapes, bare ground is everywhere. Even in places where we have plants, such as beds of upright shrubs, bare soil often exists underneath them. Still more interesting is what happens if you leave these gardens alone: voluntary plants quickly fill any open gaps and establish the same dense ground cover we see in the wild.

This tells us that the problem of bare soil is not just aesthetic, but functional as well. Every bare spot is an available niche and in the wild, every niche is filled with plants. If we do not fill these places, weeds will—requiring heavy labor or, even worse, chemical treatments to control. Mulch is one of the more benign ways to control weeds and cover the ground, but it is often expensive and can limit the potential of our plantings.

⌄ Any space around the base of plants is a niche waiting to be filled. Even low plants like *Sporobolus heterolepis* benefit from being under-planted with creeping plants like barren strawberry (*Geum fragarioides*).

⌄ The bases of trees, where traditional planting often piles volcanoes of mulch, can be filled with plants. Gray's sedge (*Carex grayi*) inhabits a pocket of soil.

Commercial landscapes designed by contractors are particularly notorious for sparse plantings surrounded by a sea of mulch. Over-mulching can create a buildup of organic matter, creating soil that is much richer than many plants prefer. Adding new layers of mulch every spring keeps our plantings in a perpetual establishment phase. It preserves bare ground, preventing other compatible species from establishing.

The alternative to mulch is green mulch; that is, plants themselves. By planting additional species to occupy the open areas, we create a lush, year-round ground cover that reduces weed invasion. A community-based approach to covering the ground is very different than the use of traditional ground covers such as ivy or periwinkle. The conventional choice is often a lone, highly aggressive species that limits biodiversity. In a designed plant community, instead of a single aggressive plant, a rich mosaic of low, compatible perennials and grasses is employed.

The essence of a plant community is the layering of different species, not only side by side, but also one on top of another. This can be achieved by vertically layering species to inhabit different niches in space and time. An obvious example is spring bulbs planted between perennials underneath deciduous trees. Even the herbaceous layer itself can be heavily layered with shade-tolerant, ground-covering plants, medium-height clumping species, and taller species with more transparent, leafless upper stems. The chapter on design process will describe how to vertically layer plantings in more detail.

For designers accustomed to creating designs in plan view, representing multiple layers of planting is often difficult. Drawings representing planting beds filled with circles and hatches may look full, but in reality, there are often large areas of bare soil underneath shrubs and trees. In fact, the graphic techniques used by landscape architects often encourage the large masses of single species that typify their plantings. Designing planting not just in plan view but in section or perspective as well encourages more thoughtful engagement about how plantings can be layered vertically. In addition, a more diagrammatic style of plan that shows how different populations of plants can overlap can help designers work with multiple layers of plants.

The ground is only one of several vertical layers that plant communities support, but we emphasize it here because it is often the one missing in traditional plantings. Our focus on covering the ground places emphasis on certain kinds of plants over others. It prioritizes those whose shapes and habits help them cover soil more efficiently. These plants tend to be low and clonal-spreading species. They are often shade tolerant because they grow under other plants. They are not always the most floriferous plants, but they are the workhorses of designed plant communities. Density is created not by cramming plants closely together, but by layering a composition vertically with plants inhabiting different spaces based on their forms.

Deeply rooted meadow species like *Panicum virgatum* and *Asclepias tuberosa* are hard to cultivate on shallow, unirrigated green roofs (bottom). If the ground cover niche is filled with a dense layer of sedum, however, a plant community can thrive (below).

Filling All Niches with Plants

—

Many traditional plantings have unfilled niches and much open soil, allowing sunlight to directly reach the ground. This is a problem because it can raise soil temperature dramatically and lead to quick evaporation of plant-essential soil moisture. Exposed soils are harsh environments for more demanding species. Think of green roof plantings and the extreme conditions plants are exposed to there. If roof media is exposed, it dries out quickly and surface temperatures can reach 160 degrees Fahrenheit and higher. Many species do not survive such extreme conditions if planted in traditional masses or spaced too far apart. However, the microclimate and growing conditions improve significantly if gaps between taller species are filled with tough ground-covering plants such as sedums. The filled niche creates better growing conditions for more demanding species—where the latter alone failed, surface temperatures stay low and moisture remains accessible for plant roots. Filling the ground niche benefits the entire planting.

Traditional planting plans show full ground cover in plan view. If the planting is shown in a section cut or perspective, the open soil and small amount of direct ground cover become visible.

Plan view: looks like dense ground cover

Section cut: reveals bare soil

Perspective: shows extent of bare soil

PRINCIPLE 4: MAKE IT ATTRACTIVE AND LEGIBLE

Much of the Western world has inherited a concept of naturalness that is tied to an eighteenth-century British concept of the picturesque. Our preference for long views, open landscapes, clean edges, and just a touch of mystery has influenced all aspects of our built landscapes. As a result, the general public has very little tolerance for wild, illegible landscapes and plantings, particularly in our towns and cities. When people encounter highly mixed plantings, they are often reminded of abandoned fields or derelict industrial sites, places often associated with urban decay or neglect.

Our reactions to natural landscapes are not just culturally conditioned, they are innate, biological responses as well. While our cultural bias for tidy landscapes often limits the potential of ecological planting, our biological responses to natural landscapes may expand them. Environmental psychologists have long theorized that our partiality for certain landscapes is based in part on their ability to provide our basic needs such as shelter and food. Multiple studies have hypothesized that the human preference for savanna—an easily recognized, productive landscape—has influenced our overuse of the English landscape style in the form of lawns. But turf and trees are not the only ways savannas can be interpreted. They can also be recreated as ecologically valuable, attractive plantings. It may be that other types of landscapes that have some of the same characteristics as savannas (traits such as legibility, openness, mystery) may work equally well as design inspiration for designed plant communities. For designers, starting with an attractive reference community as an inspiration for composed planting is an important way of creating designs that the public accepts as beautiful.

As much as we wish social conventions of landscape beauty would broaden to include a more naturalistic style, we are realists. Ultimately, the burden rests on the designer to translate ecological function into an aesthetic form. There are essentially two ways to do this. First, designed plant communities can be patterned and stylized in a way that makes them understandable, ordered, and attractive. They need not replicate nature in order to capture its spirit. For this reason, a designed plant community should be a distillation of a wild plant community, emphasizing its essential layers and patterns. Certain species and arrays may be exaggerated to create a more attractive design; other elements of wild plantings may be deleted altogether. Highly random mixes may be set against strongly massed blocks. The artful interpretation of naturally occurring mixes can create plantings that are more than the sum of their parts.

The second way to make a layered planting more appealing and coherent is to give it an orderly frame. This was first articulated by landscape architect Joan Iverson Nassauer in an essay entitled "Messy Ecosystems, Orderly Frames" (1998). The idea is that high-functioning ecological landscapes can appear messy, particularly in urban and suburban contexts. This is a problem: what is good in terms of ecological function is often disorderly, and what is neat and tidy (lawns and clipped hedges) is often not sustainable. The rich biodiversity that naturalistic planting often strives for is the very factor that

The textures and colors of a natural woodland glade (top) are conjured in Luciano Guibbelei's design for the 2014 Laurent-Perrier (bottom). Plants include *Lupinus* 'Chandelier', *Phlox divaricata* 'Clouds of Perfume', *Rodgersia aesculifolia*, and *Iris sibirica*.

Lawns are a common feature in residential landscapes (top). Their widespread use may be evolved from a preference for open savanna landscapes (bottom), which offered abundant hunting grounds and wide views.

The design by Pashek Associates for the rooftop of the David Lawrence Convention Center in Pittsburgh (top left) uses several techniques to make a highly mixed native meadow look orderly. The height of the meadow is restrained by employing low species (bottom left); short species in the front and stable, tall species toward the back prevent messy edges and floppy appearance (top right). More important, the contrast of the mixed meadow with the hardscape and monolithic sedum planting creates a balance of intricate and simple (bottom right).

Spring ephemerals randomly intermingle in this natural forest floor.

people mistake for sloppiness. Nassauer advocates that ecological landscapes should use "cues to care," that is, hints in a landscape that signal care, maintenance, and intention.

The truth is, even an artfully interpreted wild plant community may appear messy in certain contexts. There may even be moments in which a designed plant community has what writer Noel Kingsbury calls a "bad hair day." There are a range of techniques that designers can use to make a designed plant community fit in any number of sites, from small formal gardens to expansive office parks, from town squares to highway medians. One of the essential concepts of orderly frames is to surround disheveled planting with neat frames. For example, placing a mown lawn verge against the edge of a meadow, surrounding a mixed planting with a clipped hedge, or containing untidiness with hardscape elements like fences, paths, or walls.

Oehme, van Sweden's artfully interpreted forest floor
simplifies some of the diversity, uses the most floriferous
species, and amplifies the patterns to create noticeable drifts.

The importance of public acceptance for a planting cannot be understated; one
that is perceived to be beautiful will more likely be accepted, tended, and even imitated.
On the other hand, a planting deemed to be unkempt or unattractive will result in it
being ignored at best, or actively resisted at worst. One of the stated goals of naturalistic
planting, after all, is to evoke pleasurable associations of wildness, not to create discom-
fort or confusion. "When ecological function is framed by cultural language, it is not
obliterated or covered up or compromised," writes Nassauer. "It is set up for viewing, so
that people can see it in a new way."

< When plants grow as a community, as in this garden designed by James Golden, plants are not individually maintained; the entire planting is managed.

PRINCIPLE 5: MANAGEMENT, NOT MAINTENANCE

On summer weekends in many suburban settings, the din of lawn mowers and leaf blowers is a common background noise. The chorus of gas-powered machines is testimony to a landscape ideal that resists change. The same desires shape our cherished maintenance practices for planting beds. The ritualistic additions of mulch, the frequent need to prune and deadhead perennials, and regular watering are all actions meant to freeze a collection of plants in a moment in time.

Designed plant communities provide a radical departure from traditional maintenance. When plants are compatible with each other and the site, maintenance of individual species is no longer necessary—instead, the entire community is managed. This new perspective is fundamentally rooted in a design shift (Principle 1): when plants are arranged individualistically, they require individualistic maintenance; community-based planting requires management of the entire community. No longer is it necessary to give one set of plants more water and another set more fertilizer. No longer are our actions aimed at keeping one plant alive. Instead, one set of actions is applied to all populations, in order to preserve the community itself.

A management approach eliminates some of the most time- and resource-intensive maintenance activities. Watering, mulching, spraying, pruning, and leaf litter removal are generally avoided, particularly once plants reach establishment. Instead, large-scale actions such as mowing, burning, selective removal, or selective additions are used to conserve the planting's structure. The emphasis shifts toward preserving the integrity of the plant community, including its essential functional layers and its balance of species.

In many ways, the shift from maintenance to management is an affirmation that design does not happen solely during the initial act of creation. Any gardener knows that design is not a one-time gesture; it is a series of decisions made throughout the life of the planting. Design cannot be separated from gardening; it is an extension of it. Every human intervention is a decision that changes the course of a planting.

Management allows change. Because designed plant communities are dynamic, management works with a range of natural processes such as competition, succession, and disturbance. Plants that die are not necessarily replaced, but new ones are allowed to fill the gap. Plants move around, self-seed, and to a certain extent, outcompete other plants. A fundamental principle of this management approach is that when plants themselves are allowed to follow their own destiny—to self-design their own community—more robust plantings will follow.

The shift from maintenance to management requires more humility from us as designers. We need to recognize that plant communities are complex, adaptive systems shaped by their interaction with a site and other plants. In this way, management requires a series of small-scale interventions—slight adjustments of the ship's rudder—to preserve the character of a planting. Left alone, most plant communities will thrive; however, at a certain point, plantings on their own will shift enough to lose the original design intent. So management is necessary to keep the ground covered, to preserve the aesthetic quality of the planting, or to prevent aggressive plants from dominating their more demure counterparts. The manager works more like a referee than a prison guard, correcting course only when necessary.

This process must be shaped by design goals. Part of the problem with traditional horticultural maintenance is that it's made up of fixed activities that occur, regardless of whether those actions are truly needed. Mulching, pest control, pruning, and watering take place on a schedule, often without evaluating whether these activities are indeed necessary. Management, by contrast, is goal driven. Interventions take place only to steer a planting toward the design goals. Many of those goals will last throughout the project. Goals may be aesthetic, such as wanting to keep a high number of flowering species in a mix. They may be functional, as in wanting to maintain a dense ground cover to keep weeds away. And objectives may vary, depending upon the age and establishment period of the planting. Plantings in highly visible, public locations may require heavier interventions; those in more naturalistic settings may need less. How much change is allowed is the key question that will shape the design goals.

. . .

Taken together, these five principles frame a bold alternative to traditional planting. The time is right for this shift in thinking. Planting design in the twenty-first century marks a new era of increased expectations about the role of plants. Designers face more pressure than ever to create plantings that not only look good, but also perform some kind of environmental function. We need plantings to filter our storm water, sequester pollutants and carbon, cool urban temperatures, and provide habitat. Further complicating these expectations is the reality that many of our clients cannot maintain complicated plantings. Public landscapes often lack the budget or staff for maintenance, and over-committed homeowners simply do not have enough time or knowledge. These realities shift the burden to designers to create plantings that meet these seemingly unrealistic expectations, despite smaller budgets and resources.

To meet these challenges, we must design differently. We need a new set of tools and techniques rooted in the way plants naturally interact with a site and each other. This requires a deeper understanding of plants and their dynamics. The next part will look at the inspiration of nature and distill key lessons that can shape a new era of planting design. If we pay attention, plants themselves can show us the way.

> Prairie stalwart *Dalea purpurea* is a leguminous plant whose deep taproot and slender shape enable it to happily intermingle with other low grasses and forbs.

THE INSPIRATION OF THE WILD

The experience of natural landscapes is both physical and emotional. The sensory event of moving through a forest, of brushing against branches and emerging into a clearing, is enhanced by our mental associations of wilderness. Some of the best convergences of natural and cultural associations have a fairy tale quality to them: the dark folds of an Appalachian cove forest could be a setting for a Grimm brothers tale, and the gothic sublimity of a maritime live oak forest looks like a Jurassic playground. The appeal of these places is not just in their particular regional expression, but in the feeling of recognition that each of these landscapes exudes: a moment of the universal, magnified through a specific landscape.

OUR WILD HEARTS

Understanding our emotional connection to plants and landscapes holds tremendous potential for all those who design or garden. Plants can trigger emotional responses in two ways: through personal memories and through more subconscious, shared memories of common landscape patterns. Our emphasis here is on the latter. Personal memories of plants can be very powerful, but these design responses tend to be highly subjective. The scent of orange blossoms may recall a winter afternoon spent in a garden conservatory. Or a large oak tree may remind you of a special place of your childhood. These are poignant connections to plants, people, and places, but they are not always easy to replicate, particularly for public sites with multiple users. Our focus goes beyond personal memories, to a different, deeper kind of association a planting may trigger: a collective memory of nature.

While emotions are fundamentally subjective, we all share common evolutionary responses to our environment. Imagine walking down a path that bends behind a dark, contorted thicket. What do you feel? Fear? Caution? Perhaps even a tinge of curiosity? The emotions may not be exactly the same as someone else's, but they share similar characteristics. Think about the experience of hiking to the top of a mountain and looking out over the vista. The pleasant feeling of scenery was described by British geographer Jay Appleton in his prospect-refuge theory, pointing out that we have a natural preference for environments we can easily see and navigate.

While psychologists have long established the idea that there is an evolutionary basis for preferring certain landscapes, few have extended that logic to the microscale of planting design. Think about it: our ancestors spent thousands of years navigating through fields and forests. They had an intimate connection to plants. Plants helped them navigate their environments, treat their wounds, and feed themselves. Knowing how to distinguish between an edible and nonedible plant was a matter of life and death. While we may no longer rely on plants like our ancestors did, we still retain the vestiges of memory and emotion. The exact recollection may be gone, but we still have the primitive circuitry that produces emotions in response to our perceptions of safety or opportunity. When we see a certain plant or group of plants, it can create an emotional response within us, the feeling of a larger, natural landscape.

Low grass may remind us of a wide open, sunny space, making us feel expansive or uplifted. Big leaves may remind us of someplace wet, lush, and summery, such as a bottomland forest. The association between a plant and the landscape it suggests is often intuitive. We don't need a degree in vegetative ecology to understand the connection of large-leafed foliage with wet areas, or leafless succulents with dry landscapes; we sense these relationships even before we understand them. This innate recognition of a plant's connection to place explains why certain combinations feel jarring while others feel harmonious. When American landscape contractors set out herbaceous plantings in perfect rows, these artificial linear patterns can feel forced—like crops in a field. If the plants were spaced in drifts and loose aggregations, as they occur naturally, it would elicit a sense of time, of plants settling into and moving around in a place.

A massing of prairie dropseed (*Sporobolus heterolepis*) in a design by Adam Woodruff (top) recalls the feeling of a larger grassland, such as an Illinois prairie (bottom), where prairie dropseed forms a matrix for other forbs.

Individual plants of skunk cabbage (*Symplocarpus foetidus*) are reminiscent of floodplain meadows (top); a sea of the large-leafed plants conjures a sense of dampness, shade, and forest understory (bottom).

Perennials are layered by different soil depths in this dish garden. The pink splashes of diamorpha (*Diamorpha smallii*) hug the upper rims of the depression, while hairy groundsel (*Packera tomentosa*) fills the center.

Perennials, rushes, grasses, and meadow flowers enhance a water feature designed by Sarah Price for her 2012 Telegraph Garden, a wonderful interpretation of the mineral-rich upland streams and rills of North Wales and Dartmoor.

The gradient of a hillside is expressed in horizontal patterns of different forbs and grasses such as common reed (*Phragmites*) in the mid-ground and little bluestem (*Schizachyrium scoparium*) on drier slopes in the distance.

The precise emotion elicited by a planting is less important than creating a moment of engagement. People may have multiple, complex, and often contradictory emotions within a single landscape. A dark woodland path may feel foreboding to some and beckoning to others. What we see in both of these responses is a moment of resonance, a pulling out of oneself to encounter a landscape directly. As designers, we cannot control what people will feel. But we can set the stage for these encounters. In fact, the layering of emotions is what makes some landscapes compelling visit after visit. Understanding how to exploit the emotional associations can elevate planting design from the merely decorative to a meaningful art form. And if people connect with a landscape, they are more likely to invest in and care for it.

LANDSCAPE ARCHETYPES

To elicit an emotional response with planting, we must create patterns that people recognize. So it is the elusive essence of a wild community that we seek to find and reinterpret. What complicates this task is that there are thousands of plant communities around the world. Focusing on each individually could take a lifetime and, more important, distract from rather than clarify our task. After all, a montane oak-hickory forest of Virginia may be meaningless to someone in southern England, but a forest is a concept that both will understand. In order to create plantings with emotional resonance, we must first start with a point of reference that has broad appeal.

These reference communities are archetypal landscapes. Archetypes refer to a collectively inherited concept, a sort of universal prototype from which other, more specific models are derived. When applied to landscapes, an archetype refers to the distilled essence of the place, those most basic and memorable patterns of vegetation. A forest might be deciduous or coniferous, tropical or temperate, dry or wet. While these regional and climactic variations certainly matter, the goal is to set these aside long enough to understand the essential layers that connect all forests.

Our focus on archetypes as the inspiration for design is important because it describes the connection between a physical plant community and our feelings, memories, and associations. It is this confluence of real landscapes overlaid with our emotional experiences that should inspire our own planting designs, helping us to translate plant combinations into emotional experiences, not just ornamental arrangements. Focusing on archetypes also allows us to develop a universally applicable method of designing and planting. So many resources on native plants are regionally focused, limiting their appeal to a narrow audience. By understanding the essential patterns and dynamics of a forest or a grassland, one can then take that knowledge and apply it regionally. These classic landscapes give designers the flexibility to create plantings responsive to the needs of the client or the site.

Planting inspired by the wild is best when nature is interpreted rather than imitated. In many ways, our love for wild places can be our worst enemy, too often distracting us with its exquisite plants and delightful complexity. We cannot see the forest for

The open feeling of this Colorado meadow makes it naturally appealing.

the *Trillium grandiflorum*. As a result, the rote imitation of native plant communities advocated as a means to create more natural landscapes too often produces poor facsimiles of the originals. Increasingly, people are turning to their gardens as a form of habitat restoration—a goal we share—but in the process of creating a haven for native plants, the spirit of the original landscape is lost. The problem is literalism; merely importing the right plants is not enough. We must re-create the patterns and framework that gives those plantings context. Archetypes encourage us to do just that.

In order to describe a design process that has international appeal, this section covers three archetypal landscape communities: grasslands, woodlands and shrublands, and forests. Despite the myriad communities represented by these categories, each archetypal community is composed of simple layers of plants that perform specific functions. We have chosen three archetypes that are relevant to most of the temperate world.

70

With the exception of Antarctica, grasslands exist on every continent in the world. In North America, they are known as prairies. In South America, they are pampas. In eastern Europe and Asia, they are called steppes. And in Africa, they are savannas or veldts. Even areas where temperate woodlands are dominant, grasslands exist in scattered pockets of meadows, glades, and mountain balds.

GRASSLANDS

Grasslands exist between forests and deserts, often in the drier interiors of continents. Two environmental factors shape grasslands: the low average rainfall, and regular external disturbances such as fire, grazing, mowing, or avalanches in high altitudes. These environments are generally moist enough to sustain deep-rooted grasses, but dry enough to prevent forests from easily establishing. In general, dry grasslands are shorter, whereas wet grasslands are taller and can be dominated by aggressive, clonal species. While the species composition of grasslands is often naturally generated, their continued existence frequently depends upon external disturbances.

THE EXPERIENCE OF GRASSLANDS

Few other archetypes have captured the imagination of plant enthusiasts and designers in the last half century like grasslands. In many ways, grasslands provide the ideal mix of design features: at a large scale, they are grand and uniform, giving them tremendous emotional impact; yet on a small scale, they are intricate and layered, creating a staggering diversity of flora and fauna. Several naturalistic planting movements such as the New American Garden, the New Perennial movement, and German mixed perennial plantings (*Staudenmischpflanzung*) have drawn deep inspiration from naturally occurring

Low-height grasslands such as this meadow are highly appealing for their legible forms and long views.

This heroic landscape is a grassy bald, a rare high-elevation grassland of the southern Appalachians. Balds are unusual in that, unlike alpine meadows whose lack of trees is generally caused by cold temperatures, they are warm enough to support trees. Many theories exist for their presence, but their precise origin remains a mystery.

grasslands. It is easy to understand their contemporary appeal: grasslands offer a sense of openness and freedom in a world increasingly walled in by urbanization. The image of an ocean of grasses moving in the breeze—with its overwhelming sense of space and sky—continues to seduce a generation of designers.

The classic image of grassland (at least, our idealized version of it) is low in height. This allows us to see over it. Our preference for highly legible landscapes like this probably evolved from a desire to identify threats (predators, invading armies) from a distance. Long vistas have a soothing effect on us. This simple fact has shaped much of the history of landscape design, as framed vistas over grazed pastures were a central feature of the English picturesque movement.

Viewed from a distance, the entire tapestry of grasslands blends into a monolithic green background. Only when these species flower or fruit do they create thematic seasonal displays and patterns of color and texture. During these moments, grasslands can look as if a painter splayed streams of color across a landscape. Yet from a distance, a grassland's thematic species melt into larger drifts of color and texture. The hard edges and details are lost. The linear form of many accent forbs and low woody shrubs within the matrix of grasses is often a response to subtle changes in moisture and elevation. Subtle creases in topography (where water collects in a field) are often visible with distinct changes in vegetation. The simplicity and clarity of these linear forms is what we find so appealing.

Worldwide, grassland plant communities have countless expressions which are a direct result of their responses to climate conditions, underlying soils, hydrology, and the frequency and nature of disturbances. Low-height grasslands can be found in dry, moist, and wet soils.

As a result of these varied conditions, grasslands have unique, site-dependent colors and textures which are consistent within a plant community. For example, species that make up dry grassland communities often have thick cuticles or leaf hairs to minimize transpiration. These adaptations make their leaves appear blue, grey, and silver-green. Their very fine leaf morphology reduces surface area and therefore water transpiration. On the other hand, lush and broad leaf textures are typical in hydric grassland communities. They do not need to

DRY GRASSLAND COMMUNITY

Grassland communities have different color ranges depending on their environment. These color ranges are created through different leaf morphologies and colors. While leaves in dry habitats appear blue or silver-grey, leaves in moist to wet environments are shiny and deep green.

COLOR RANGES OF GREEN

conserve water. In order to take up nutrients diluted in water, plants have to transpire large quantities of water through their foliage. Leaves of hydric grassland species have some of the deepest, most saturated, and lushest shades of green. Protective cuticles are not necessary in their consistently wet environment.

One of the unique features of native grasslands is the visual similarity of textures and shades of green. As a result, these communities have a strong sense of legibility and authenticity. Harmonious ranges of color and texture signal that a planting has evolved with the site over many decades. The slow process of competition and evolution results in a rich relationship of plants to place, producing plants that have similar colors and

MOIST GRASSLAND COMMUNITY

WET GRASSLAND COMMUNITY

textures to each other and the site itself. We may perceive this harmony subconsciously, but we are not always aware of it. Some planting feels dissonant to us, and a difference in color and texture is often the cause. For example, introduced exotic vegetation often has a slightly different color range than the native vegetation. Many invasive species such as Japanese honeysuckle (*Lonicera japonica*) or cheatgrass (*Bromus tectorum*) outcompete natives because they either stay evergreen throughout the year or emerge as annuals earlier than the native warm season grasses. In spring, the distinction between the bright green introduced plants and the predominantly dormant native vegetation is apparent.

The range of colors in this autumn meadow is harmonious with the forest backdrop, adding a feeling of authenticity. Drifts of early goldenrod (*Solidago juncea*) dominate the foreground while little bluestem (*Schizachyrium scoparium*) and *Andropogon virginicus* make a striking mass in the mid-ground.

Even the sharp orange blooms of monarch flower (*Asclepias tuberosa*) do little to dim the predominance of green grasses in the summer.

Joe Pye weed creates a thematic seasonal display in this grassland.

Golden ragwort (*Packera aurea*) forms large colonies of low basal foliage that tightly cling to the ground much of the year, but in the spring they shoot up spectacular, long-blooming yellow flowers.

ESSENTIAL LAYERS

The visual essence of all archetypal grasslands is the horizontal line. Grasslands lack tall vertical vegetation structure such as trees and shrubs. The tallest vertical elements are tall grasses and forbs. Short shrubs are present in a few cases, but they generally mingle with perennials at the same height. The limited vertical structure does not mean that grasslands are sparsely vegetated—in fact, the opposite is true. Wherever conditions are favorable, the horizontal stratum is very dense and does not leave a spot of soil uncovered. Grassland plant communities have more species per square meter than many forests. How is that possible?

Close examination reveals that the one horizontal stratum is not homogenous at all. Instead, it consists of many substrata or layers, making grassland plant communities as complex as any other plant association on earth. These layers are present above- and belowground in the root zone. Like most plant communities, the stratified composition of grasslands is the result of ecological niche dynamics—a way for many species to limit competition. Different stem and root morphologies allow species to grow side by side without directly competing with each other. The incredible morphological diversity allows plants to extract nutrients and water from different belowground horizons, and air and sunlight from different aboveground horizons. Tall species can be found in

Irises, dandelions, and golden pea (*Thermopsis montana*) create linear patterns in this subalpine meadow, due to subtle differences in soil moisture.

the exact spot as ground-covering perennials. One draws light and air from four *inches* above the soil, the other from four *feet* above it. Soil profiles show that the same morphological diversity happens underground as well. Deep taproots grow right through shallow, fibrous root systems, allowing species to share nutrients and water from different locations.

These communities are layered, but the exact boundaries of those layers are fluid and overlapping. To keep it simple and be able to translate this complex structure into design principles later, we divide plant communities into layers most applicable to designers: the visible upper stratum which we call the "design" layers, and lower, more ground-covering mantles, called the "functional" layers. These layers are not traditional ecological categories, since plants here have various adaptations and strategies; instead, we have categorized them in ways most relevant to designers.

> Even early in the season, the large basal foliage of cup plant (*Silphium terebinthinaceum*) acts as a structural plant among the prairie dropseed (*Sporobolus heterolepis*) and white prairie clover (*Dalea candida*).

>> Structural grassland plants tend to hold their form even in winter. Here a clump of wild bergamot (*Monarda fistulosa*) is interspersed with blazing star (*Liatris* sp.).

Structural layer

The structural layer is formed by tall forbs and grasses. Plants in this category are called structural because, unlike bulbs or ephemeral species, their shapes persist for large portions of the year. Most stems are strong enough to persist through the winter and are essential for a planting's winter interest. They create the backbone of a plant community. Examples include switchgrass (*Panicum virgatum*), Indian grass (*Sorghastrum nutans*), rattlesnake master (*Eryngium yuccifolium*), cup plant (*Silphium perfoliatum*), Joe Pye weed (*Eutrochium fistulosum*), and big bluestem (*Andropogon gerardii*). Some plants like goldenrod (*Solidago* sp.) or blazing star (*Liatris* sp.) can form seasonal flowering displays, but it is really their structural qualities that most contribute to their presence in grasslands. Many of these plants are highly competitive and survive by towering

DESIGN AND FUNCTIONAL LAYERS

Showy species of the grassland's upper design layer are the plants with which designers are most familiar and are used to create patterns of color and texture. Underneath, however, are species of high functional value. They often stay hidden under taller and showier plants and quietly perform essential erosion control, soil building, and weed suppression.

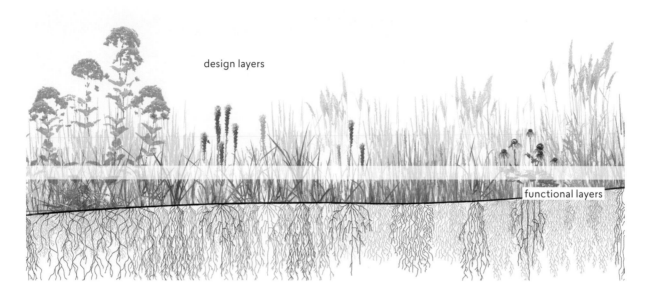

design layers

functional layers

STRUCTURAL LAYER

The structural layer of grassland communities is built by tall species that tower over smaller plants.

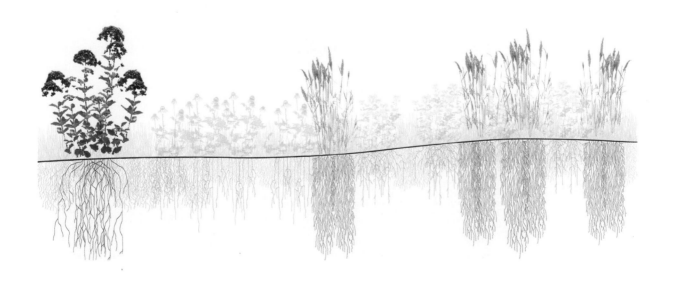

Eutrochium fistulosum, Sorghastrum nutans 'Sioux Blue', and *Senna hebecarpa* (left to right) are all excellent structural plants.

Eutrochium fistulosum, Sorghastrum nutans 'Sioux Blue', and *Senna hebecarpa* (left to right) are all excellent structural plants.

over shorter species. Their stems are thicker and stronger in order to carry the weight of a taller plant. Plants in this layer vary in how closely or tightly they group. Some such as switchgrass (*Panicum* sp.) form individual clumps, while others like Joe Pye weed (*Eutrochium* sp.) are clonal and form groups or larger drifts that create patterns visible even from a distance.

Long-lived and clumping perennials and grasses of this category are valuable design elements. Their reliability and mannered behavior make them excellent visual anchors. While other parts of a planting change through the year, structural plants keep a design visible and stable for many years to come. Structural perennials emerge at the same time as other perennials, and they need several months to reach their final height. They never occur in monocultures in the wild. Just like canopy trees in a forest, structural perennials have few leaves in the lower strata of their plant community, allowing shorter species to grow directly under them and cover their lower stems.

Seasonal theme layer

Plants in this layer create seasonal themes with their flowers or texture. These perennials occur in large quantities within grasslands, so when they flower or fruit, they visually dominate for a few days or weeks. After blooming, they melt in with other green plants. Grassland plant communities typically go through several waves of color per year. Some occur regularly. Moist meadows, for example, have an intense purple event, when sweeps of New York ironweed (*Vernonia noveboracensis*) flower every fall. The recurrence and repetition of these plants helps to stabilize a grassland community visually,

∨ Species of *Eutrochium* and *Solidago* glow in the morning sun, enhancing the seasonal theme in this meadow.

⩔ Nigel Dunnett and Sarah Price's design for the European Gardens of London's Olympic Park is a stylized version of summer hay meadows. Here, species of *Leucanthemum*, *Deschampsia*, and *Sanguisorba* form dramatic seasonal themes.

Sweeps of *Vernonia nove-boracensis* form stunning purple themes in August and September.

giving it a sense of order and legibility among the diversity. Plants in this layer tend to be longer-lived, including perennials such as daisies (*Leucanthemum*), primroses (*Primula*), daylilies (*Hemerocallis*), buttercups (*Ranunculus*), St. John's wort (*Hypericum*), salvia, and irises. The layer may also include showy grasses like tufted hair grass (*Deschampsia cespitosa*), little bluestem (*Schizachyrium scoparium*), and splitbeard bluestem (*Andropogon ternarius*). Some seasonal flowering events are weather-dependent or triggered by fires. For example, many desert annuals such as California poppies (*Eschscholzia californica*) and African daisies (*Gazania*) flower after rare rain events.

The work of University of Sheffield professors Nigel Dunnett and James Hitchmough is worthy of mention here. Both have exploited the extraordinary design potential of seasonal theme plants in grassland plantings. Nigel Dunnett's concept of pictorial meadows uses seeded annuals sown with other perennials, for sweeping blocks of color throughout the growing season. The perennial plantings of the gardens in what is now Queen Elizabeth II Olympic Park in London also relied heavily on the dramatic character of a seasonal theme layer for spectacular waves of color. The strong aesthetic character of their carefully designed mixes increases public acceptance of naturalistic vegetation. In Germany, landscape architect Heiner Luz artfully uses seasonal themes to create stunning effects in planting.

These plants all form strong seasonal themes at different times of year: (left to right) *Asclepias purpurascens*, *Callirhoe involucrata*, and *Monarda bradburiana*.

SEASONAL THEME LAYER

Seasonal theme plants create color or texture effects in plant communities at certain times of year. Many of these species are popular garden plants because of their showy flowers and attractive foliage.

Ground cover layer

The plants of this layer are functional in a design sense because they hug the ground, prevent erosion, and suppress weeds. In grassland communities, this layer is formed by a carpet of species belonging to genera such as *Carex*, *Packera*, and *Viola*. Many ground-covering plants are rhizomatous or stoloniferous, which allows them to quickly move around taller species and fill any gaps they can find. A few plants in this layer are legumes—such as the genera *Desmodium*, *Lespedeza*, and *Oxytropis*—giving them the ability to fix nitrogen from the air in the soil. Other ground covers self-seed and have the ability to colonize gaps within plant communities. These ground-holding qualities make this layer essential for design.

Plants in this layer adjust to different levels of sunlight through the year. In the spring and early summer, plants receive full sun. Later in the year, as taller perennials grow above the ground cover layer, these plants are partially or fully shaded and sheltered from sun. This can cause some species to go partially dormant, to survive the deep shade of summer and fall. They often flower and fruit before this happens, and use the available growing window similarly to spring ephemerals in forest plant communities. Some geophytes such as *Triteleia* and *Crocus* fall in this category. Their large underground storage organs allow them to survive in the wild during unfavorable growing conditions.

In the early spring, the ground cover layer is clearly visible. Its diverse textures and colors become the main design elements before taller perennials take over the show. *Packera aurea* blooms among the budding foliage of *Physostegia virginiana* and *Deschampsia cespitosa*.

With the possible exception of bulbs, species of this layer are often not very floriferous or showy. Many are grasses or leafy perennials that do not produce spectacular flowers. From a design point of view, this is not a problem as these low plants are often not very visible; the chief problem is that so few of them are commercially available. For example, path rush (*Juncus tenuis*), an outstanding and adaptable ground-covering rush, would not turn many heads on a garden center shelf—part of the reason why this layer is frequently missing in man-made plantings. Attempts to replace this essential layer with mulch frequently fail.

In designed landscapes, the top soil horizon is often highly disturbed and compacted, which can lead to poor infiltration and surface runoff. Ground cover species can help break the compaction up and heal this soil over time. As their often-shallow root systems regrow every spring, these species break through compacted soil strata and enrich the soil with organic matter.

Juncus tenuis, Carex amphibola, and *Deschampsia cespitosa* 'Goldtau' (left to right) are three examples of unspectacular workhorses of the ground layer.

GROUND COVER LAYER

Ground-covering plants occupy the lower layer of grassland communities and weave in wherever enough sunlight reaches the ground. Their root systems are generally shallower and do not directly compete with deeper roots of taller species.

Lobelia cardinalis, Ipomopsis rubra, and *Eschscholzia californica* (left to right). Dynamic filler plants occur in all habitats. *Lobelia* prefers moist to wet conditions, while *Ipomopsis* and *Eschscholzia* thrive on dry sites.

Dynamic filler layer

Grasslands are the home of highly dynamic and opportunistic species. These plants are less competitive than taller perennials and grasses and are commonly shorter lived. Their strength lies in the ability to find and inhabit gaps within grassland communities. Annuals, biennials, and short-lived perennials produce large amounts of seed, which quickly move across the land. Wherever plant cover opens up, seeds of these dynamic plants find optimal conditions to germinate. As they grow, they form seed which is stored in the soil, remaining viable for many years to come.

Filler plants have great value in the early stages of plantings. Their fast growth and ability to flower and fruit within their first or second year help stabilize new plantings and cover soil quickly with intended species. Once dynamic species have reached the end of their life, slower-growing but longer-lived perennials can take their place.

The other layer: time

Grasslands are formed of species with different metabolisms and life cycles. No species is always present. Some go dormant at different times of year and others reach the end of their lives and disappear. In other words, species constantly fade in and out, occupying or freeing up space. For example, cool season species actively grow in early spring. But as soon as temperatures reach a certain height, these grasses go summer dormant. Their metabolism does not allow them to photosynthesize during hot temperatures. Summer dormancy allows them to survive unfavorable times. When this happens, species with warm season metabolisms occupy nearly the same spot within a plant community. Warm season species are acclimatized to the heat of summer and fill now-available

< Golden Alexander (*Zizia aurea*) and Columbine (*Aquilegia canadensis*) light up this *Carex*-dominated planting bed. Columbine is relatively short-lived, but also readily self-seeds, opportunistically filling gaps.

Packera aurea growing directly underneath tall Joe Pye weed (below). Both create better growing conditions for each other; tall Joe Pye weed shades *P. aurea* in summer. In winter, *P. aurea* covers the soil underneath (bottom), keeping it insulated.

Plants that Work in Several Layers
—

Some grassland species fall into several categories, depending on their companions and the overall height of the plant community. For example, blazing star (*Liatris spicata*) may form the structural backbone of a two-foot-tall meadow community. However, it can also create seasonal color and texture themes in an eight-foot-tall meadow community with tall big bluestem (*Andropogon gerardii*) and great coneflower (*Rudbeckia maxima*) as structural elements. Not all plant categories are always present; they may also occur at different percentages. Some grassland plant communities may have very few structural species and can be harder to read. Others have well-developed structural backbones that create fantastic winter interest.

Species that visually dominate a meadow in late spring (left) will eventually be covered by taller species. In the same meadow a few months later (right), *Baptisia australis* and *Tradesantia ohiensis* endure the heat in the shade of *Coreopsis tripteris* and *Silphium perfoliatum*.

niches until their productive cycle is completed by fall. Through temporal layering, no space is ever empty and soil is never bare. This is why high species diversity is possible even on small footprints.

In fact, within grassland plant communities, some species could not survive without their temporal companion species. For example, golden ragwort (*Packera aurea*) occupies a wet meadow's ground layer in spring. It goes summer dormant in late June, right after it sets seed. Without taller perennials shading it through the heat of summer, *P. aurea* would probably not survive the drier conditions and higher sun intensity of July and August. By mid-June, structural perennials like Joe Pye weed (*Eutrochium fistulosum*) and ironweed (*Vernonia noveboracensis*) start growing into higher strata, providing essential shade for *P. aurea* and other species of the ground layer. Both plants grow in the same spot without ever directly competing with one another.

Layering plants not just in space but also in time is one of the most important inspirations nature gives us for better planting design. It results in highly functional landscapes. Dense temporal layering of species can reduce management costs. Soil is always densely covered and weeds will find fewer places to grow. In addition, this strategy produces more biomass for pollutant and nutrient sequestration, creates denser root systems for storm water treatment, and provides better erosion control and soil building function. The consistent plant cover created through temporal layering provides stable habitat, food, and cover for all forms of life.

California poppy (*Eschscholzia californica*) is a classic example of a plant that emerges only when conditions are favorable. Like many desert annuals, its seed sits dormant, waiting for annual rain events which trigger an explosion of color.

PROBLEMS TO AVOID IN GRASSLAND-INSPIRED DESIGNS

Our focus on archetypal grasslands emphasizes the features of these wild plant communities that most people prefer. But when grasslands vary from these norms, they often become less appealing. These problems are important for designers to understand and avoid. They include:

Towering plants. The human preference for easily navigated landscapes makes meadows above eye level intimidating, particularly in urban contexts. These taller grasslands are only acceptable if viewed from a distance. Using plants that are generally lower than waist height is one way of making grassland-inspired planting more appealing and acceptable for small residential landscapes or urban parks.

Too little visual interest. If grasslands are visually dominated by grasses, they can feel monotonous, empty, and boring. Not having enough flowers or rich textural elements can leave a planting feeling too grassy. Color matters, so be sure to exploit the rich potential of the seasonal thematic layer to give your plantings several waves of color through the year.

Jarring combinations. Clashing blends of color and texture can occur if species from different habitats or parts of the world are mixed, losing a potentially wonderful harmony. Try to avoid mixing in plants from communities other than grasslands. Species adapted to non-grassland habitats will almost always stand out, creating a planting that lacks internal coherence and congruence with a site.

Woodlands and shrublands are scattered all over the world, often situated between grasslands and forests. They are often composed of widely spaced trees with a mixture of shrubs and grasses on the ground plane. The larger, more iconic examples are found on the west coast of continents in the midlatitudes, in places that have Mediterranean-like climates. They are also known as chaparral, matorral, cerrado, and scrubland. But smaller woodlands and shrublands are also scattered throughout the interior of continents, including montane zones in windswept areas above the tree line. These landscapes generally receive more rain than deserts and grasslands, but less than forests. The unpredictability of rain is a key characteristic of such biomes. Plants have adjusted to the cycles of dry summers and moist winters.

The finely textured species of the ground cover layer have adapted to dry soil conditions and high sun intensity. Shrubs and trees usually show the same morphologies and have thin leaves or needles. The matching textures of ground and canopy strata give this archetypal landscape consistency and harmony. Plants in woodlands and scrublands have developed many of the coping features of desert plants to help them survive the hot, dry summers. Plants like sage bushes have small, needle-like leaves that help to conserve water. Some species have leaves with waxy coatings or leaves that reflect sunlight. Many are annuals which flower after spring rains and survive during the dry summer as dormant seed.

The lack of consistent tree canopy is the single defining element of woodlands and shrublands. Trees are usually shorter than those in forests, due to the lack of rainfall or infertile soil. Soil and climate conditions support tree cover, but disturbance is high and prevents the formation of permanent forests. Fire shapes the character of many of these ecoregions; numerous plants have responded to frequent burning with the development of underground roots in the case of grasses or sage bushes, or thick bark in the case of scrub oaks, pines, and cork trees. Open woodlands and shrublands are home to an astonishing number of plant and animal species. These habitats include a wide diversity of light conditions and microclimates. They contain fragments of forest edges and open grassland, allowing the landscape to support species from both ecosystems.

WOODLANDS AND SHRUBLANDS

THE EXPERIENCE

Archetypal woodlands combine the visual clarity of low-height grasslands with the shelter provided by scattered trees or shrubs. Their visually open expressions are highly appealing. The introduction of low shrubs and trees adds a hint of complexity and mystery, yet the wide spacing and open nature of the vegetation keeps this complexity from overwhelming.

The experience of moving through these landscapes is rhythmic. Scattered trees and shrubs create a mosaic of human-scaled spaces. As you walk through woodlands, room-like enclosures open into wide expanses, which close again into dense, shrubby vegetation. It is a landscape of contrast: openings and closings, alternating light and

< The striking patterns of trees, shrubs, low woody plants, ferns, and grasses create a remarkably readable plant community.

The savanna-like quality of many woodlands makes an ideal inspiration for suburban-scale landscapes, mimicking the patterns of trees, hedges, and lawn.

Even in early winter, the stratified nature of woodland plant communities—found in the lines between grasses, shrubs, and trees—is distinct.

dark, warm and cool, sun and shade. Light conditions change quickly; our eyes do not have enough time to adjust. This is why light areas feel brighter and dark areas feel darker than forests. Unlike the vast cathedrals of forests, clustered trees create cloistered rooms with long views over the open landscape. The low trees make rooms feel secret and sheltered.

In contrast to the monolithic character of grasslands, the vegetation of woodlands and shrublands is highly patterned. The edges between grassy fields; drifts of low, woody vegetation; and tightly clustered trees tend to be distinct. Often these zones express changes in soil depth or moisture. The patterns often have a lyrical quality to them: long melodic drifts of grasses are framed by harmonic repetition of shrub masses and tree clusters. The layered structure prevents visual clutter and lends clarity to these archetypal landscapes.

While contemporary planting design gravitates toward grasslands as inspiration, woodlands and shrublands are underutilized points of reference. They are vegetative models of human-scaled spaces, making them one of the more engaging landscapes to be in. The grand scale of grasslands and forests can make them difficult to translate for small urban and suburban landscapes. But the room-like quality of woodlands and shrublands—particularly the tight tree clusters adjacent to open fields—works remarkably well for suburban contexts, suggestive of the way planting and paths might frame open lawns. Another underutilized inspiration is the clearly patterned character of the vegetation. These patterns are created by highly contrasting mixes of grasses, shrubs, and trees. While each layer of these ecosystems is composed of many sub-layers, they look visually distinct.

Trees are often clustered in woodlands and shrublands, creating varied patterns of open and closed vegetation.

Trees, low woody shrubs, perennials, and grasses form clearly delineated zones on this woodland floor. In summer, green vegetation dominates, but by fall and winter, the different layers become visible.

ESSENTIAL LAYERS OF WOODLANDS AND SHRUBLANDS

In addition to horizontal herbaceous vegetation, open woodlands and shrublands have vertical elements such as trees and shrubs. This archetypal landscape has two visually dominant layers: a mostly herbaceous ground cover stratum and a taller, clustered canopy stratum. The latter is formed by individual specimens, clumps, or small groups of trees or shrubs. Since shade is not nearly as deep as in forests, plants grow year-round even under dense canopy. The available sunlight extends the growing season for herbaceous species and they are not limited to spring ephemeral survival strategies as in summer-dark forests. Spring ephemeral species can be present in open woodlands but they are usually not as abundant as in forest plant communities.

The woodland ground stratum is as complex as in grassland communities. However, the diversity of microclimates created by scattered trees and shrubs adds higher seasonal diversity. Some sheltered areas show the first new leaves of spring much earlier

A longleaf pine savanna illustrates the simplicity of the canopy layer often associated with woodlands, balanced by increased diversity on the ground layers.

than exposed areas. Where grasslands form large and uniform communities, woodlands create patches of vegetation on a much smaller scale. For example, some spots of shaded ground layer may still be frozen in the shade of a shrub facing north, while on the southern side of the same shrub one might see the first delicate flowers of spring.

Canopy layer

A small number of species takes the visual lead within the canopy, giving this landscape its unique feel and sense of place. Repetition of a handful of signature tree species is perhaps the single most defining element of woodlands and shrublands. Pine canopies, for example, suggest serpentine barrens or loblolly savannas, whereas scrub oaks may recall a montane woodland.

The strong competition between canopy species of dense forests causes trees to stretch and grow toward the light. In this process, trees lose some of their unique growth habits. In open woodlands, however, trees and shrubs are not directly surrounded by others of the same species. The space between allows them to grow into "wolf trees," which expand into unique forms and silhouettes. Imagine wind-battered pines or a stand of majestic oaks in a lush pasture or hayfield: an arboretum of wonderfully diverse tree and shrub forms.

Woody layer

In many woodlands and shrublands, low woody vegetation actually functions like an herbaceous ground cover. In fact, the distinction between woody and herbaceous plants tends to blur in this layer. Plants like Russian sage (*Perovskia*), sagebush (*Artemisia*), or bluebeard (*Caryopteris*) are often classified as both herbaceous and woody. These are often low and spread clonally. For example, in many ericaceous plant communities such as heaths, moors, and peatlands, low woody plants of the genera *Calluna*, *Vaccinium*, and *Andromeda* form dense masses. In drier shrublands like the California chaparral or the *matas* of Portugal, dwarf subshrubs in the genera *Juniperus*, *Artemisia*, and *Rosmarinus* dominate the ground plane. All of these plants are typically acclimated to low fertility soils and arid conditions. Many shrubs in this layer are deeply taprooted.

Saw palmetto dominates the woody layer of this longleaf pine community.

100

The diversity of herbaceous plants in a woodland can be as high as an open grassland. Here the fall colors of grasses, asters, goldenrod, and milkweed echo the foliage of the trees beyond.

The open character of a woodland canopy allows grasses to dominate the ground plane. *Chasmanthium latifolium* makes a dense ground cover in this river barren.

Harsh site conditions as well as frequent fires shape the vertical layers of woodland plant communities. Twigs, thin branches, and especially limbs close to the ground fall victim to fires. This process prevents dense understory and keeps the woodland open. Tree trunks are often black from fires and charcoaled stumps can be found scattered across many of these landscapes. The entire plant community is adapted to fire. Trees and shrubs may lose some of their limbs but readily resprout after fires. For example, pitch pine has very thick bark and the ability to resprout from its main trunk after fire. Herbaceous species push new foliage up from their sheltered crowns or deep root systems. Some species survive only as seeds in the ground.

Herbaceous layer

The herbaceous layer includes plant species that tolerate wide ranges of sunlight conditions ranging from dappled shade to full sun. The space under trees and shrubs is normally densely covered by shade-tolerant plant communities. Just like grassland communities, the herbaceous layer consists of a ground cover stratum and taller structural layers. Structural forbs and grasses are only a small part of the vertical structure of woodlands and shrublands. The more dominant vertical elements are trees and shrubs, which create structural frames of the archetype, even in winter. Therefore, structural perennials and grasses feel less visually important than taller woody species.

Japanese honeysuckle (*Lonicera japonica*), Russian olive (*Elaeagnus angustifolia*), and other invasive vegetation choke this woodland, rendering its layers illegible.

While most of the herbaceous layer looks like a single consolidated expression, occasionally the composition of the herbaceous layer will shift, particularly when faced with the dense shade of tree clusters. These pockets of intense shade will create conditions similar to many forests, welcoming more shade-tolerant sedges and perennials.

PROBLEMS TO AVOID

What makes woodlands so appealing is their rich layering, the clean lines between different plant types, and the interplay of open and closed vegetation. When this structure is lost, woodlands can look more like dense thickets—closed and unapproachable. Maintaining the clarity of the structure is critical.

Blurred layers

Shrublands and woodlands include inherently uncomfortable expressions that remind us of unmanaged land and failed garden projects. Often these uncomfortable forms recall landscapes in transition, particularly highly mixed, early successional landscapes. Examples include overgrown farm fields transitioning into early woodlands and unmanaged parks or gardens smothered with invasive vines and trees.

Woodland and shrubland plant communities become less appealing to visitors when the vegetation is highly mixed. Woodlands are composed of mixed height vegetation, and while this can be an asset, it can also obstruct views and confuse one's perception of a landscape. This is particularly a problem when the ground plane is overly complex. Dense understories covered in tangled vines, overgrown shrubs, and tree seedlings are difficult to navigate and feel threatening to us. Exotic species can be a problem in any plant community, but the diversity of vegetation types in woodlands offers a wide range of habitats in which invasives can get established, requiring added management and even surgical interventions. Large-scale techniques like mowing and burning can help, but they also may encourage tree and shrub seedlings to grow more prolifically, resulting in taller, less-open planting. The alternative is to more clearly separate layers of trees, shrubs, and herbaceous plants into distinct zones, each of which is then vertically layered with compatible species.

Traditional planting techniques such as massing species together or repeating a handful of thematic plants can help distinguish layers. This does not require monocultures of single species; instead, the plan may feature matrixes of a dominant plant. For example, blocks of single species such as low shrubs can still be underplanted with low sedges and ground-covering forbs to make them more diverse and resilient. As long as distinct edges between blocks are visible, the composition will read clearly. The other strategy to consider is using a palette of mostly native woodland and shrubland plant communities. Inelegantly combined species from different woodlands or archetypal landscapes may look like an artificial hodgepodge, disrupting the fine balance of color and texture that gives a community a feeling of authenticity.

Poor spatial composition

The most stunning natural examples of woodlands and shrublands are very balanced landscapes. There is often a pleasant proportion of taller canopy species, mid-height shrubs, and low herbaceous plant. Too few canopy species can create more of a shrub-savanna than woodland, resulting in a planting that feels more exposed than true woodlands. Too many mid-height shrubs can obstruct views, producing confusing, maze-like spaces. Trees spaced either too closely together or too far apart can make plantings that are either too dense or too barren. Clarity of spatial organization is the key. This may require careful editing of existing plantings, particularly mid-height vegetation, to effectively screen and frame usable garden rooms.

The high canopy, lack of a shrub layer, and intricate herbaceous floor make this open forest very appealing.

Forests represent one of the most abundant and ecologically complex biomes on the planet. They cover vast stretches of the temperate world, including most of eastern North America, the majority of Europe, and large swaths of Asia. Wherever soils are deep enough and the frequency of disturbance is low, forest ecosystems thrive. Forests have a wide scope of expressions, including the evergreen forests of the northern hemisphere, the deciduous forests of the midlatitudes, and the tropical forests of the equatorial middle. They range in height from scrubby coastal thickets to towering redwoods, grow in swamps and on dry soils, and vary in age from a few decades to thousands of years old.

FORESTS

A forest is a patchwork of plant communities with varying shade tolerance. Forests differ from woodlands in that they generally have a consistent evergreen or deciduous tree canopy that casts shade on the ground below. Whereas canopies in woodlands are generally wide apart, trees in forests touch, creating shade on the ground. Forest shade levels vary depending upon the canopy and time of year, creating many levels of shade density. If a forest is deciduous, the shadow levels change greatly throughout the year, gradually diminishing in autumn and increasing in late spring. Some temporary canopy openings (caused either by soil anomalies, wind, or disease) let more light reach the forest floor. This allows patches of less-shade-tolerant plant communities to thrive within the otherwise deep shade environment, increasing habitat and species diversity. Canopy openings are similar to spotlights on theater stages and can be powerful design tools. A stream of light piercing through canopy to illuminate a patch of ferns creates a dramatic chiaroscuro effect.

THE EXPERIENCE OF FORESTS

Forests evoke seemingly contradictory feelings, a testament to the complexity of meanings that make them so emotionally resonant. Forests appear as settings for our ancient myths, fairy tales, and even contemporary fiction. They are the homes of benevolent fairies and wicked witches, places both of shelter and grave danger. They are simultaneously familiar yet mysterious, nourishing yet foreboding. The dichotomy of emotions we project onto forests is perhaps a reaction to different qualities of spaces. Dense thickets of trees may appear unnavigable and menacing, whereas more open groves may have a sacred quality to them. The veneration of certain trees or groves was widespread in many cultures, perhaps a testament to the timeless appeal of forest plant communities.

What is it that makes some forests appealing and others threatening? We prefer wooded areas that have very little understory and wide spacing between tree trunks. Visually pleasing forests feel open and have a cathedral-like roof formed by tall trees.

The essence of an open forest is conveyed in the repetition of trees and the expression of the forest floor. Here a glade of birch and sedge (*Carex pensylvanica*) forms a room-like space.

Very few shrubs and tree seedlings obscure this view. The forest floor is covered in moss, lush ground covers, or sometimes just a thick layer of freshly fallen leaves. These forests are easily navigable and allow open views underneath the tree canopy. Large-sized, older trees are almost universally preferred to younger specimens, perhaps an evolved intuitive sense of the value of large trees as shelter, timber, or nourishment.

Archetypal forests create a dark and shady atmosphere with a unique microclimate that affects us in many positive ways. Dense canopies provide shelter from wind and sun, making them cooler in the summer and milder in the winter. Forest ecosystems make us breathe deeper and reduce our heart rates. The air has been filtered by forest vegetation and smells mossy and earthy. Not by accident, historic healing gardens were inspired by open forests. Also not by coincidence, etheric oils of pine, fir, and cedar are used in bathing salts, aroma therapies, and massage oils. In addition to its therapeutic qualities, a hike in an open forest engages our senses and sharpens our attention. We hear distant bird songs, branches swaying in the wind, footsteps rustling in leaves. We smell moss, fungi, and ferns after a good, soaking rain. We see the fireworks of warm colors on crisp autumn days. This sensory immersion engages a more primordial experience of self, offering a sense of integration with the natural world too seldom felt in today's fast-paced world.

ESSENTIAL LAYERS OF FORESTS

Vertical lines define forests. Upon entering a forest, our eyes automatically follow these lines up from the bases of trees and shrubs into the forest canopy. Little sky shines through this canopy; during the summer it often stays completely hidden behind glittering green leaves. Forests reach taller heights than any of the aforementioned archetypes and therefore have the richest diversity of plant layers. A forest is not just an accumulation of individual trees. Plants of a forest are linked with each other in ways we are just now beginning to understand. They communicate with one another, nurse small seedlings through a network of mycorrhizal fungi, and wage chemical warfare against other species.

Closed tree canopy

The canopy layer is created by trees that grow to the light as straight and fast as they can. As soon as they reach canopy height, they spread out in order to expose their foliage to as much sunlight as possible. Canopy trees have almost no lower limbs; the majority of their leaves are at the top. In fire ecologies, canopy trees have thick bark to protect themselves from low-heat ground fires. Many trees produce fruit irregularly to lower the risk

Newly emerging foliage of sugar maple (*Acer saccharum*) in spring.

Redbuds (*Cercis canadensis*) form a seasonal thematic layer at the edge of a hemlock forest. A limited understory creates the open character of vegetation at eye level that makes archetypal forests so appealing.

of all seeds being consumed by predators or destroyed by extreme environmental conditions. Canopies are often referred to as ceilings and function as such in many functional ways. They control not only the light coming through, but also the movement of air. Trees adjust to varying light levels by producing thinner, shade-tolerant leaves during their youth on the dark forest floor, then thicker, waxier, sun-tolerant leaves when they grow higher.

Patchy or absent understory and shrub layer

Just under the dense canopy, we find the open forest's sparse understory layer. It is formed of scattered groups of shade-tolerant shrubs, small trees, and young canopy tree seedlings. In order to be successful under limited sunlight conditions, these species adjust their seasonal growth cycles to the light windows created by the leafing-out of canopy trees. They grow in the more sheltered horizons closer to the ground and are protected from frost. Because of this sheltering, shrubs and trees of the understory can afford to flower and produce leaves before canopy trees. For example, spicebush (*Lindera benzoin*)

leafs out long before the canopies of oak, ash, and maple, using available sunlight to produce flowers and seeds during the early growing season. In the shady summer months, spicebush and many others make the most of the little available sunlight by adding massive amounts of chlorophyll to their leaves. This makes their foliage take on very dark shades of green, which helps sustain them in low-light environments.

The understory is sheltered from wind and sun. Humidity is usually much higher than at the canopy level, and the combination of deep shade and high humidity results in large leaves. One of the most striking examples of this phenomenon is the leaf of the empress tree (*Paulownia tomentosa*). When young, it can be as large as sixteen inches across. Mature, canopy-level trees, however, have leaves with diameters of only six inches. This adaptation strategy is essential for canopy species. They have to be very shade tolerant when small in order to survive in the ground layer and understory. There they wait until a canopy tree falls, and a spot opens up—an opportunity that can take many years. Before then, tree seedlings are exposed to many dangers: being browsed by deer, burnt by fires, or broken off by snow, ice, and falling branches. Their ability to resprout after disturbance is their secret to success. Northern red oak (*Quercus rubra*) seedlings can be decades old and have extensive root systems, while being only a few inches tall. As soon as a canopy tree falls and enough sunlight is available, the small oak seedling shoots toward the light. The deep root system it grew for so many years is now a significant advantage in winning the race for the forest canopy.

The low height of many forest floor perennials makes it easy to mix and mingle them without a messy look. Here, wild ginger (*Asarum canadense*) is teamed with foamflower (*Tiarella cordifolia*).

Herbaceous ground cover layer

Perhaps the most striking element of archetypal forests is their ground layer. It is easy to admire the soft mix of colorful bulbs and ephemerals, forgetting how brutal this natural habitat really is. In order to survive under dense shade and compete with the massive root systems of trees, plants of the ground cover layer have evolved impressive morphological and life cycle adaptions. To survive, herbaceous species become masters of timing. Like understory plants, spring ephemerals leaf out and flower long before the canopy layer closes and creates deep shade. Archetypal forests are famous for their dense carpets of spring ephemerals that bloom in early to late spring, such as the extensive blankets of Virginia bluebells (*Mertensia virginica*), trillium (*Trillium grandiflorum*), and mayapples (*Podophyllum peltatum*). Large underground storage organs make this life cycle possible. Spring geophytes, such as in the genera *Erythronium*, *Crocus*, and *Narcissus*, have bulbs that store resources needed to bridge long periods of dormancy and sprout again the next spring. Spring ephemerals go completely dormant after they have completed their life cycle in mid- to late spring.

The ground layer also includes species that only go partially dormant in summer. Just like spring ephemerals, their main growing season is limited to spring. However, after they complete flowering and seed production, they keep their leaves and persist in the ground layer until frost. Creeping phlox (*Phlox stolonifera*) and spotted geranium (*Geranium maculatum*) are perfect examples of this strategy. Their leaves are broad and deep green. This leaf morphology allows them to perform photosynthesis in deep shade and create necessary energy for the coming growing season, which spans the entire summer and early fall months.

The third life cycle adaptation is that of conservative growth. Species such as tree groundpine (*Lycopodium dendroideum*), Christmas fern (*Polystichum acrostichoides*), and mountain laurel (*Kalmia latifolia*) are highly adjusted to deep shade environments. Their leaves contain large amounts of chlorophyll and they can perform photosynthesis in very heavy shade. Instead of producing new foliage every season, species with this survival strategy are mostly evergreen, allowing them to save energy and valuable resources.

⌃ A dense mat of trillium pools at the base of an oak. May-apples poke through.

∧ The herbaceous layer is itself richly layered with plants of different morphological adaptations. Here mayapple (*Podophyllum peltatum*) mixes with other ephemerals, such as Virginia bluebells (*Mertensia virginica*), cutleaf toothwort (*Cardamine concatenata*), and trillium.

⌃ Allegheny spurge (*Pachysandra procumbens*) adopts a conservative growth strategy, focusing on slowly spreading under limited light conditions.

∧ Bunchberry (*Cornus canadensis*) is a subshrub ground cover that thrives in moist, acidic soils.

Woodland floor perennials are masters of timing. Some, like the *Phlox divaricata* and trilliums seen here, bloom in spring then go dormant during the heat of summer. Others like Christmas fern (*Polystichum acrostichoides*) seen in the back persist year-round.

Species of the ground stratum are subject to strong pressure from shallow tree root systems. Some herbaceous plants avoid this competition altogether by means of even shallower root systems. They evolved to root mainly in the duff layer—the shallow pockets of leaf litter and surface soil between tree roots. *Oxalis acetosella* is a good example of this morphological root adaptation.

The other layer: time

Site conditions in the lower strata of deciduous forests change dramatically during the year. Plants that thrive in the full sun of spring are often not able to tolerate the deep shade of summer. Spring ephemerals are succeeded by more shade-tolerant species such as ferns and sedges that emerge and cover soil until the canopy becomes more translucent again in fall or spring of the following year. The temporal sequencing of species is an essential element of stable plant communities, providing consistent habitat, ecosystem function, and an attractive plant cover. Interestingly, even under such difficult conditions, soil within healthy forests is almost always densely covered with plants. In fact, the diversity of light conditions actually increases the diversity of species within forests.

Archetypal forests have stunning seasonal color. Starting with a dense mat of spring ephemerals, forests often glow with spring color long before the region's last frost date. The summer months are dominated by the color green. Flowers are rare, as shade-loving ferns and sedges cover the ground. Fall belongs to asters, goldenrods, and woodland sunflowers. By winter, the ground layer is covered by fallen leaves, leaving only evergreen ferns and sedges poking through their winter blanket.

PROBLEMS TO AVOID

Dense, tangled thickets, contorted paths, and impenetrable vines can stir feelings of unease. In order to achieve desired responses to planting design, we must be aware of what makes forests feel intimidating and less appealing.

Obstructed views

Forests can have thick understories which restrict views and make navigating difficult. Forest floors invaded with thick shrubs and strangling vines create ecologically and visually distressed plant communities. Multiflora rose (*Rosa multiflora*) and Japanese honeysuckle (*Lonicera japonica*) can transform open forests into dense jungles of impenetrable greenery. Disturbed and young forests with less-dense canopies allow more sunlight to reach the ground and are therefore more susceptible to more profuse ground vegetation. Until forest canopies are dense enough to shade out unwanted ground vegetation, management is needed. Some of the dense understory has to manually be thinned out or burned to restore the health and sensory cues of an archetypal open forest.

Dense, unnavigable vegetation makes this coastal forest less appealing. The predominance of eye-level vegetation in the midstory erodes the clarity of the space.

Missing layers

Stability is only achieved if all elements of a forest are present. However, environmental issues as well as design mistakes often lead to missing layers within forest plant communities. For example, an overpopulation of deer can deplete plants of the ground and understory layers. The resulting lack of tree seedlings can interrupt the formation of the next generation of canopy trees. The other problem with too few species in the ground layer is that it creates gaps and open soil—an invitation for often very deer-resistant exotic species.

Mixing plants from different habitats

Combining species from various forest types can create assemblages that lack the harmony of archetypal forests. If trees are too visually different, they will never form a harmonious forest canopy. Even many of our so-called natural forests are either planted or enhanced by man. In many cases, trees for timber production end up growing next to volunteer species that seeded in on their own. For example, planted spruces, pines, and oaks frequently mingle with volunteer hickories, black gums, and cherries—associations that may have never evolved together naturally. City parks are often even more exaggerated examples, assemblies that look and feel entirely man-made.

The edges of all three of the classic landscapes we've discussed have beauty and distinct patterns of their own. While edges are not a specific landscape type, their pervasiveness in urban and suburban areas makes them worth special mention here. Edges occur both naturally and as a result of disturbance. We want to emphasize the patterns, layering, and depth of naturally created edges, over the more sharp lines of human-created edges.

EDGES

In many ways, the natural areas within cities and suburbs are predominantly edge landscapes. Our heavy use of land has pushed natural areas into a vast web of linear strips: narrow slivers of trees, shrubby undergrowth along drainage channels, and bands of herbaceous plants along parking lot edges. These natural areas rarely have the luxury of depth. The dynamic interactions of grasslands or forest, for example, often depend upon having enough depth of space to sustain certain populations of plants. Trilliums are often found only in the deep interior of forests. The narrower plant communities become, the more edge dynamics shape their species and behavior.

Edges are the result of changes in site conditions. Some of these changes are natural, such as the presence of a lake next to a forest or the recession of trees above a timberline. Others are man-made, as with an agricultural field bordering a forest. In the wild, conditions rarely change abruptly, gradually transitioning from one type to another. For example, the soil along a water's edge is a gradient: on the edge of a lake, soil is entirely saturated, but as it moves up the bank it is moist, and higher still it is dry. The edge of an herbaceous plant community gradually emerges from standing water to a dry

meadow. Another example is the woodland edge. Fire or windfall may have disturbed a forest community, and now a meadow grows where trees used to thrive. Edges can be stable or in flux. The herbaceous clearing within the woodland may eventually revert back to trees if succession is allowed to take place. Stable edges generally occur next to man-made structures such as highways and golf courses, or barriers in the natural landscape such as lakes or rock formations.

Edges are where species from different landscapes overlap. The result is high species diversity. For example, grassland species often stretch into the bright edges of woodlands and woodland shrubs often seed into edges of meadows. It is not unusual to see deep forest ephemerals, such as *Podophyllum peltatum* and *Dicentra eximia*, on the edges of woodlands. The fact that so many species thrive in edge communities makes them especially valuable and worthy of good management.

Natural edges are wide and tapered, gradually transitioning in height from one landscape to another.

A well-developed edge contains plants that transition in height between the two adjoining landscapes. Competition for sunlight shapes much of this dynamic. Lower species gradually taper up to the highest point, creating a feathered edge between different plant communities. What is important about this gentle tapering is that it results in a stable, "sealed" edge, protecting the interior of ecosystems from disturbances that might threaten them. This gradual feathering of plants is too often lost in built landscapes. Our desire to maximize space results in sheared, thin edges that often expose sunlight and bare soil to parts of a landscape not adjusted to these conditions. Think of the forest clearings that make room for highways and new housing developments. After construction is finished, new infrastructure is generally surrounded by sharp, unstable edges. Forest interior trees with tall, naked trunks sharply border turf and parking lots. Tree trunks that have never been exposed to direct UV light now face full sun. As

Oaks along the edge of this forest loosen up and feather
into an open grassland.

a result, many invasive species thrive in these edges. Mowers along the sides of roads
spread seeds of plants like garlic mustard (*Alliaria petiolata*) and Japanese stilt grass
(*Microstegium vimineum*), which then invade forest understories from the edges. Therefore, edges require more of our attention and higher management input than the more
stable landscape interiors.

Edges are often far from stable, but the opportunities for planting design are huge.
Enhancing an edge by adding back layers to simulate the natural feathering of stable,
natural edges increases the evocative quality of a landscape and makes it feel more
authentic. Well-designed edges improve and stabilize microclimates and growing conditions for plants. Such edges create healthier and more resilient plant communities,
which in turn reduce management cost. They also strongly benefit a huge number of
other creatures that share the planet with us.

Rock outcroppings support herbaceous species when pockets of soil are deep enough. Here *Helianthus porteri*, an annual endemic to Piedmont rock outcroppings, blooms in early autumn.

GRASSLAND EDGES

The most appealing grassland edges are neat and of low height. If bordering a lake or stream, for example, they can be abrupt and only a few feet wide. Other edges gradually ebb out and change in height. This is often caused by increasingly inhospitable soil, causing stunted and dwarfed plants due to lack of water and nutrients or extreme salt concentrations. For example, meadow communities often gradually ebb out into rock outcrops or coastal dunes.

Not all grassland edges are equally appealing. Some are tall and more like a wall that divides one landscape from another. For example, a wall of eight-foot-tall common reed on a water's edge or a dense stand of *Panicum virgatum* where a storm water basin meets a parking lot are not likely to be perceived as attractive. In fact, the upright shape of these grasses is an adaptation to help the plant reach for light in the center of plantings. Instead of transitioning a grassland archetype into another landscape, tall vegetation acts more as a barrier between landscapes. In the wild, grasses and forbs right on the edges of plant communities have leaves closer to the ground than if they were growing in the center of a meadow. Many edge species have a gracefully arching habit, which allows them to cover bare ground more fully than upright grasses, and makes them ideal plants for edges. These are excellent frame species that we can use later in the design process to create orderly borders around planting.

WOODLAND AND FOREST EDGES

When woodlands or forests border ponds or meadows, unique edge communities result. Trees and shrubs do not only grow up into the canopy, they also grow sideways into the open space over water or grassland. Trees are generally smaller toward the edges than in the center of a forest or woodland. Exposure to wind as well as plenty of sunlight coming from the sides is the reason for shorter height. Shrubs are a particularly important part of woodland and forest edges. Early successional species from genera such as *Aronia*, *Cornus*, *Myrica*, *Baccharis*, *Clethra*, and *Sassafras* provide dense nesting cover and food for a wide range of fauna. Woodland edges are home to gorgeous large-flowered forbs, such as *Echinacea purpurea* and *Helianthus divaricatus*. They too often stretch toward the light, giving woodland edge communities an interesting dynamic. It appears as though these species want to leave the shade and venture out into the open. Uncomfortable woodland edges are dense and full of brambles, poison ivy, vines, and invasive species like multiflora rose. They are so thick with vegetation that views into the woods or meadow are limited to a few glimpses.

Because edges are so ubiquitous in built landscapes, their design potential is tremendous. Connecting and expanding these fragments can greatly increase their beauty and ecological function. Fingers of woodland edges, for example, may be connected with larger woodland stands. These edges also serve as buffer zones between developed areas. They provide visual screening and noise mitigation, as well as filtering storm water and pollutants.

· · ·

Archetypal landscapes are a way of understanding the intersection of actual plant communities and our emotional perception of them. By looking more closely at the layers of some of the most powerful landscapes, we can extract the essence of these communities, merging their visual and ecological layers into a single, universal expression. This simplification is not meant to reduce the true complexity and endless variation of naturally occurring plant communities. In fact, creating regional variations that authentically capture the flavor of local plant communities is our goal. But to do this, we must first distill and amplify the essential layers designers can use to make these archetypes meaningful in urban and suburban landscapes.

THE DESIGN PROCESS

Great planting design is the result of three harmonious interactions: the relationships of 1) plants to place, 2) plants to people, and 3) plants to other plants.

The first relationship describes the deeply rooted connection plants have to their sites. When a plant merges into its environment, all of its best qualities are magnified. A pocket of ferns growing in the crevice of a mountain boulder testifies to the slow process of building soil from raw rock. Just as important is a plant's ability to engage and enchant. Watching a forest floor erupt with trilliums and foamflower—like all great seasonal moments—is a delight not only for our eyes, but also our spirits. However, it is a plant's relationship with its botanical companions that gives a garden its potency. The dark silhouette of perennial seed heads set against misty grasses reminds us that when plants are paired with their right companions, planting can be more than the sum of its parts.

HONORING THE THREE ESSENTIAL RELATIONSHIPS

Lackluster planting, like so many things, is often the result of an imbalance in relationships. Conventional landscaping, for example, has traditionally focused on relating plants to people, often ignoring the context. Consider the heavy use of chartreuse or plum-colored foliage shrubs in front of commercial developments. Each plant is bred for a pleasing color, yet the assembly of high-contrast shrubs is often jarring to the eye. Conversely, ecological planting has focused heavily on relating plant to place, sometimes at the expense of human pleasure. You may have stumbled upon a sign at a schoolyard butterfly garden explaining the benefits of natives, only to wonder where the garden actually was.

Our method honors and balances all three. We start with understanding how plants relate to their place, developing an intuitive process of observing and analyzing a site. The goal of this process is to translate a site to its archetypal form, providing an inspiration that connects plants with our emotional experience of them. The next step is to develop the design framework that relates plants to people. Providing a structured frame around mixed planting helps it to relate to urban and suburban contexts. Finally, plants are related to other plants by carefully layering them into various niches, resulting in a truly functional community with the highest possible ecological value.

Before getting into details, we want to make a couple of points. First, in order to develop a method that anyone can use, we have created a simplified process. The prescriptions shared here are not meant as narrow formulas to be slavishly followed, but rather open-ended processes that encourage creativity and personalization. You will find no rigid recipes, cookie-cutter plant lists, or stylistic dictates. Instead, we offer an approach that rests on the creativity of the designer. We hope it will result in as many different styles as there are practitioners. Second, creating designed plant communities requires thoughtful engagement on the part of the designer. We believe this process is teachable and can be mastered; but to be successful, one must understand a site, develop a palette of plants, and install and manage the plantings appropriately. There are no shortcuts, no quick and easy plant lists or combinations. After all, the preassembled plant lists offered by most garden literature often prove useless outside of their regional contexts. We are aiming for something higher—a fittedness of plant to place that results in unmistakable harmony. This requires site-specific solutions, not paint-by-number planting design. But the effort is worth it. Planting design in the age of climate change will demand more from the designer than ever before. Mastering this process will reward you with plantings that are lush, layered, and more resilient.

Though not a naturally occurring combination, *Carex muskingumensis* and *Petasites japonicus* fit together well because they come from similar habitats and their shapes interlock. They radiate harmony and authenticity even to an untrained eye.

∨ *Panicum amarum* 'Dewey Blue', *Phlox paniculata* 'David', *Echinacea purpurea*, and *Pleioblastus distichus*. All originate from very different habitats and therefore have different leaf colors and textures. This composition does not look authentic.

∨ These meadow species originate from similar habitats and match in color and texture. However, poor design composition makes the planting look messy and unstructured.

In winter, landscapes can be easier to read and archetypes are more visible. Spread-out trees in snow recall open woodlands, while denser trees in the background indicate a forest archetype.

RELATING PLANTS TO PLACE

Successful planting design starts with the big-picture understanding of a site. Begin with context: where is your site located and what landscape surrounds it? Are you right on the edge of a forest that was cleared for development? Are you surrounded by open fields? Or are you in the middle of an extensive suburban woodland? This larger orientation should come first and helps clarify which archetypal design goal will feel authentic and work with surrounding landscapes. For example, if your residential project site was carved out of a dense hardwood forest, your site could be a spectacular design bringing the quality of forest or woodland to light.

Once you are aware of the larger context of a site, you are free to explore what character your site has and what archetypal landscape is hiding within. You could compare this step to the work of a sculptor asking what figure is waiting to be revealed within a block of marble. Decisions about hardscape materials, plant selection, and combinations are later built on this essential first step. Skipping this step, or not completely understanding a site before designing its details, almost always leads to unfocused and messy design.

Seeing the archetypal landscape hidden in your site is not always easy. Look past distracting details to essential forms, such as this grassland and forest edge. The clarity of the landscape is immensely appealing, an idea that must be preserved through the design process.

The skill of stepping back and seeing through the clutter, right down to the bones of a landscape, can be learned and perfected. Landscape elements that have to be removed should not be taken into account here. This may include invasive vegetation, unsafe trees, or damaged walls or fences. Site analysis should be based on elements that will most likely transition into the future landscape design. Do not be misled by details such as colorful flowers, existing pathways, and garden ornaments. Distractions are plentiful at times, but homing in on the true character of a site will lead to the right design decisions.

The end goal of this process is to identify what archetypal landscape your site wants to be. While our small urban spaces will never be true grasslands, woodlands, or forests, they can look and function like more distilled versions of them. Understanding your site in terms of an archetype allows you develop an appropriate palette of plants, and perhaps more important, to create a space that pleasurably echoes nature.

EXPLORATION BEFORE ANALYSIS

Understanding your site is vital, but we want to emphasize a different kind of analysis than is routinely advocated by landscape architects and designers. Site analysis—particularly the inventorying of natural and built systems made popular by landscape architect Ian McHarg in the 1970s—is still taught in reverential tones in landscape architectural schools across the globe. It focuses on a surficial examination of topography, hydrology, soils, and even plant communities. The idea of compiling a series of ecological inventories to understand a site is certainly sound. After all, each of these components is a piece of a larger story about a place. But in practice, many landscape architects lack the time and scientific rigor to do this effectively. Even when it is done well, the results are difficult to translate into a clear design direction. This data-driven process reduces the site to a mound of little facts that say almost nothing about the character, mood, or quality of the site—the very qualities designers are trying to enhance.

We do not want to diminish the importance of scientific site analysis. In many ways, data gathering will be the future of large-scale landscape design. However, we want to emphasize the qualitative experience of a site, not just the quantitative analysis of it. After all, great planting should delight and please us, not just serve a functional purpose. Our starting point emphasizes a different kind of investigation—one based more on the art of exploration. Most of the information we need to understand about a site is easily observed. Underneath each site is a series of natural systems, layered over with centuries of human intrusions, modifications, and buildings. The most important elements are often the most obvious ones. A steep hillside speaks hundreds of messages, from the movements of glaciers, to the patterns of drainage, to the kinds of plants that want to grow on it. Each rock is a totem of geology and erosion; each cluster of trees is a witness to the soils underneath; each shadow is a map of the sun's path. We need no laboratories or computers to understand a site. We need to go outside and walk.

To understand the messages of a site, you must first learn to explore and observe. Walk, first aimlessly, following what draws your attention. This kind of non-directed attention brings to light our intuitive responses to a site. What were you attracted to? How did you move through the site? What felt uncomfortable? These emotional responses matter as much as our intellectual analysis, often revealing the character of a landscape. Like following a divining rod, certain elements of a site will beckon, others will push us away. If you are attracted to a certain vista or enclosed spot, this may become a focal point of the final design. If an undergrowth of vines feels uncomfortable, this may be something to remove. The more clearly we can distinguish what draws us and what repels us, the easier we can separate what should stay and what can go.

Freely wandering a site can offer another gift: the power to resurrect dormant genetic or psychological traits that helped our ancestors survive in the wild. A fear of snakes or heights, for example, is not always learned, but an innate, biological reaction to threats in the landscape. Conversely, our attraction to sheltered promontories or flowers may be a response to safety or fecundity. In the modern era, there is little in most designed landscapes to either threaten or sustain us—at least in a primal sense—but these cues are still valuable for designers interested in eliciting emotional responses. The parts of a site that draw the strongest psychological reactions must be noted. Even our negative reactions to certain parts of a site are worth exploring. A dense thicket of existing shrubs may be used to line a path that opens into a sunny, low meadow. To purge a site of all its negative features is not always necessary. In fact, amplifying some of these elements can create more appealing landscapes. Most conventional landscapes seem to strive for a kind of benign pleasantness that is pretty, but ultimately boring. Engagement is the goal. It is the layering of contrasts—of dark and light, closed and open, ominous and propitious—that creates the most engaging spaces. But to create these delicious contrasts, we must first find them. The simple art of strolling, of casual indirection, can reveal the emotional touchstones of a site.

Existing vegetation with a strong structural presence such as this *Clethra alnifolia* is often a good starting point in a design. Planting that thrives and contributes to the character of the space can be amplified.

FIELDWORK: THE ART OF OBSERVATION

After experiencing a site, the next step is to observe and note the elements that give a place its character. The goal is to separate crucial, character-giving elements from nonessential ones. This kind of fieldwork is not entirely art or science, but a bit of both. Start

with noticing the larger characteristics of the space. Is the landscape dense and closed with trees and shrubs, or open and navigable? Where are the high and low points? The basic elements of tree cover, surface, and water flow are clues to finding a location's true character. Other factors are important, but at this point we are not yet analyzing soil type, hydrology, microclimate, or plants. Focusing on these details before understanding the big picture can complicate and confuse planting design.

Often, new landscapes are more easily understood than established gardens. Familiar terrain is hard to see with fresh eyes; the character often gets misinterpreted because of our emotional connections and countless memories with it. Sometimes we build our gardens around existing plants that add little real character. Other times our desire for certain kinds of plants leads to choices out of character with the site. To gain an accurate understanding of a space's essence, zoom out a bit, distance yourself, and squint your eyes. What do you see? Is it wide open and sunny like a grassland? Are there a few existing trees, widely spaced apart, as with a woodland? Or are there dense canopies more akin to a forest?

Often the hardest part of this process is deciding what should go. Clearing a site down to its essential elements is both liberating and terrifying. Our attachments to certain plants, or comfort with the status quo can make purging hard. But ruthlessness is an asset. Eliminating undesired existing vegetation opens up a fresh canvas for new planting. Getting rid of long-standing invasive species, in particular, can be cathartic. Stripping invasive vines off tree trunks or artfully pruning an overgrown shrub can reveal the beauty in existing forms. But overclearing can be problematic as well. Do not clear more than you can replant. Removals can disturb soil and open up light, inviting weeds and invasive species.

Making a blank slate of a site is often tempting for designers. It offers ultimate freedom of choice. But the problem with choice is that it induces a kind of design paralysis. We now have access to a staggering variety of plants. As a result, it is possible to have Mediterranean gardens in the mid-Atlantic, British walled gardens in Japan, and Alpine rockeries in Brazil. Choice places a burden on us to create character that was not there before. For a project with enough money and vision, this can work. Central Park in New York City was largely created this way. Many of that park's distinguishing natural features—the rolling landform, rock outcroppings, and forests—were not givens on the site, but were imported and shaped. On most sites, however, there exists some fragment of character, some existing landform or remnant vegetation that can be built upon. The point is to learn to see how the givens of a site can be its best assets. A steep slope could be viewed as a problem, or as a canvas for creating great drama, recalling the charms of a mountain walk. Deep shade can be the bane of a garden—or the defining quality of it. Wet clay can be cursed, or it can be the element that unifies a plant palette.

SPECIAL CONSIDERATIONS: HIGHLY URBAN OR DISTURBED SITES

Designers working in cities may encounter sites with little or no natural or vegetative features. The typical urban project site—rooftops, paved plazas, small parks, and sites formed from demolished buildings—offers few natural features to read and interpret. In fact, cities often have such a long history of disturbance that almost everything about them, from their soils to their existing vegetation, is artificially created. While the clues may be less obvious, the process of reading a site to identify an archetypal inspiration still works in cities. In many ways, because urban locations are so removed from forests or meadows, plantings that recall these places can be that much more pleasurable. The popularity of the High Line in New York City, for example, demonstrates the appeal of a wild, meadow-like planting amidst a backdrop of skyscrapers.

Urban sites can be read in much the same way as natural landscapes. While existing vegetation may not play a large role, other factors are relevant. The amount and intensity of sunlight can by itself determine the kind of planting that should be on a

∨ Winter reveals the essential layers of a forest in the New York Times building courtyard.

∨ HMWhite's design for the New York Times building courtyard transforms a shoebox of a space into a woodland glade.

site. For example, the intense sunlight and exposure of a building rooftop might imply a grassland or meadow community as a design goal. The soft swaying of grasses on a green roof can be a welcome alternative to more static sedum mats, evoking high-elevation meadows. Street-level sites are entirely different. A courtyard surrounded on three sides by a tall building may have light levels similar to a forest floor. The lobby courtyard for the New York Times building in Manhattan playfully uses birch trees set atop gently mounded hillocks of sedge and fern to suggest a forest floor. To achieve this effect, landscape architects HMWhite engaged specialists to measure light levels and created models using 3-D software. The end result is a small naturescape that feels expansive because it so elegantly distills a woodland glade.

Pay careful attention to the movement of sun across an urban site; in many ways, the quality of city shade is different than tree shade. Urban areas are surrounded by hard surfaces, so reflections can actually create more light in some places; in others, non-reflective surfaces can result in deeper shade than the filtered light of a tree canopy. Since urban shade is not entirely enclosed from above like tree shade, sites can have a few hours of harsh, direct midday sun. Choosing plants tolerant of a wide range of light conditions—such as many species native to forest edges and woodlands—is a safe choice for ground level sites.

In addition to light, soils also can be a strong factor in determining a goal community. Soil depth is a particular constraint in cities. Many planters are built over structures like parking garages or roofs. Six inches of soil is about the minimum you will need to establish perennials and grasses; twenty-four inches or more can be enough to establish small trees, given enough total horizontal soil volume. Limited soil volume will certainly affect plant selection. Plants with long fibrous roots, like grasses, may fare better in limited soil depth than plants with deep taproots. When soil volume is limited, irrigation becomes important. Raised planters dry out faster due to the exposed surface area and lack of groundwater. Even plants that could otherwise thrive in dry environments may need supplemental water in urban conditions.

As extreme and unnatural as urban conditions may seem, there is likely a native plant community in the wild that thrives under similar conditions. Finding these inspirations requires research, but a good match will be worth the effort. Former industrial sites often have highly compacted or polluted soils. Grassland communities are often a good fit for brownfield sites, providing deep roots that absorb heavy metals and create soil. Urban plazas might be inspired by rock outcroppings. These communities feature plants that endure punishing heat, long periods of drought, total anaerobic conditions after a rain, and small pockets of soil. Biofiltration planters might be inspired by the plants that thrive along the upper edges of open stream banks. These plants can tolerate long periods of drought, followed by scouring washouts during heavy rains. Almost any urban problem site has a parallel condition in the wild; we just have to make the connections.

The repetition of birches, the amplified topography, and the sedge- and fern-filled ground plane pack the essence of a forest into the New York Times building courtyard.

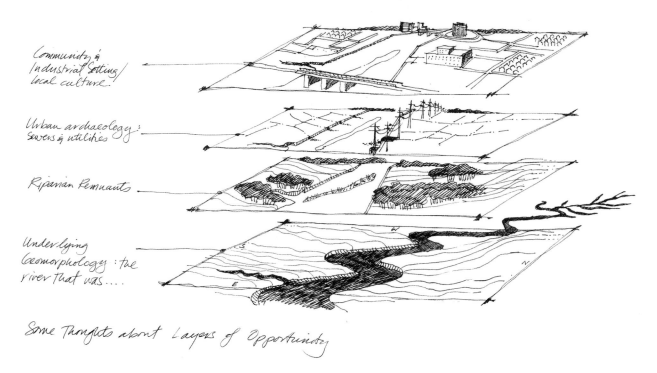

This sketch by Faye Harwell, FASLA, explores the natural and built layers of Four Mile Run stream, an industrialized tributary of the Potomac River. Drawing courtesy Rhodeside & Harwell Landscape Architects and Planners.

Community & Industrial Setting / local culture.

Urban archaeology: sewers & utilities

Riparian Remnants

Underlying Geomorphology : the river That was....

Some Thoughts about Layers of Opportunity

SKETCHING AS A WAY OF SEEING

With practice and observation, one can clearly decipher the character of a site. However, the best way to find the bones of a project site is to draw. Designers frequently sketch to come up with a conceptual idea, but it is important to use sketching even earlier in the design process to understand a site. Think of drawing as a kind of visual note-taking. Whether your sketches look like a photo-real landscape or a toddler's scribbles does not matter. The point is to engage your mind in a different way of seeing. Drawing is a kind of thinking. It forces us to see landscapes as they really are. Instead of revealing objects, drawing uncovers forms, shadows, and patterns, eliminating visual clutter. This step is worth the extra effort and will highlight aspects of your site that you might otherwise miss.

Freehand your sketches on plain paper as much as you can. Position yourself within your site or on its edges, looking in. Ideally, you should have a good view of the entire site. Start by sketching large portions of the site, or the entire site if possible; this will help you avoid distracting details. If your site is too large or there are too many obstacles, use satellite photos or bird's-eye view images available online.

Whether you draw your sketches in plan view or perspective, or both, depends on your site and what you want to document. Plan view is best for analyzing tree canopy

LANDSCAPE SELECTION KEY

This landscape key works like a plant identification key, and can help identify the archetypal landscape best suited to an existing location. It is based on site elements and the landscape visible from the site.

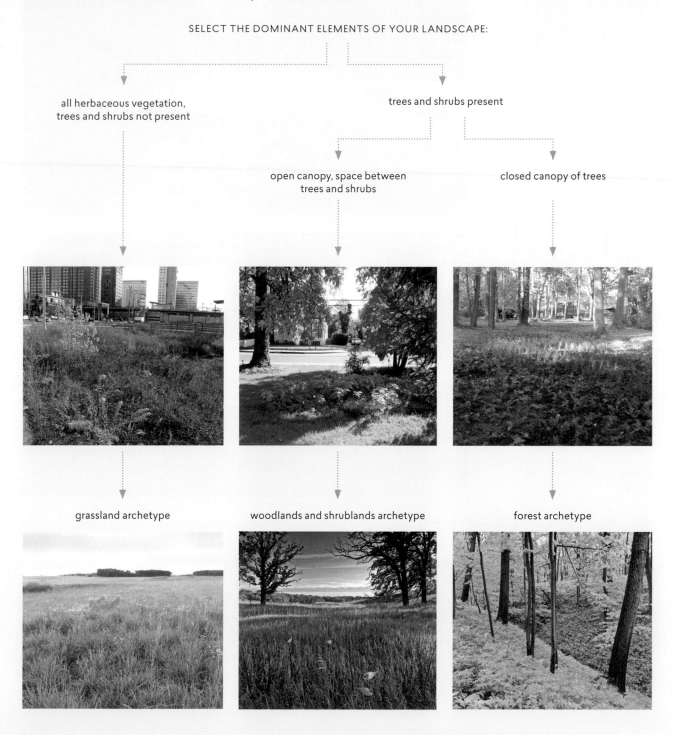

SELECT THE DOMINANT ELEMENTS OF YOUR LANDSCAPE:

all herbaceous vegetation, trees and shrubs not present

trees and shrubs present

open canopy, space between trees and shrubs

closed canopy of trees

grassland archetype

woodlands and shrublands archetype

forest archetype

A small suburban lot (top right) along a busy road hardly resembles a wild landscape, but the presence of a few large trees and the need to screen out the road make the woodland edge a fitting archetype. Two years later at the same site, a layered planting of *Rhus typhina*, *Sambucus canadensis*, *Heuchera villosa* var. *villosa*, *Deschampsia flexuosa*, and *Carex divulsa* has begun to echo a woodland edge landscape (below right).

cover and the patterns of vegetation. Drawing on top of an existing survey can help clarify scale, pattern, and the distance between elements. If you do not have a survey, try to find aerial images that show texture and can help you distinguish types of vegetation cover, such as tree canopies, rough grass, and lawn. Perspective or section sketches, on the other hand, will help identify vertical layers. While plan view flattens a site, drawing in perspective or section allows you to separate and notate the layers of vegetation. Using both methods paints a more complete picture of your site.

Sketching will reveal elements indicative of a larger archetypal landscape. For example, a few scattered trees could indicate an open woodland archetype. This element could later be picked up in your design and amplified into a forest or woodland landscape goal. Or maybe your site is located out in the open, and a meadow landscape is the logical design goal. In most cases, you will quickly get a sense for what larger landscape is hiding within an existing site. The key on page 135 will help with more cluttered and confusing sites, where landscape archetypes may be difficult to ascertain.

At the end of this first step you should have a good feeling for which landscape archetype your site could be. The next step builds on this deeper understanding of your space as we enhance elements and make the overall character of your landscape stronger and more transparent.

Gardens once were a refuge from the wild, but now we turn to them for an experience of the natural world. Our desire to have an authentic encounter with wildness seems to grow as we lose more truly wild places to urbanization. Planting can help fill this longing, immersing us not only in sensory delight, but also in a home-coming with nature. The way a grass moves in the wind or a seed head glows when backlit by the sun provides a window to a world beyond our boxed-in cities.

RELATING PLANTS TO PEOPLE

But in order for planting to recall a memory of nature, patterns of the wild must be translated into a horticultural language that relates to the structured environments of our towns and cities. Literally replicating native plant communities in urban landscapes can lead to disappointing results. Examples abound of well-intentioned rain gardens, pollinator gardens, or native restorations that look straggly and forlorn. Our small urban sites lack the advantages of scale, context, and time that their wild counterparts have. Consequently, the natural patterns and palettes that are so potent in nature must be distilled, selected, and amplified in human landscapes.

This requires creating a strong design framework for your planting. Frames provide a basic structure for the planting that underlies and supports it. They offer visual cues both within and around, to help people see and appreciate the important layers. In this section, we will talk about two frames:

- The conceptual frame of the planting itself, defined by its design goals
- The physical frame around a planting, defined by its edges

CONCEPTUAL FRAMEWORK: CHOOSING A GOAL LANDSCAPE

A designed plant community is held together by an idea. It is the concept that first shapes the planting, even before the plants themselves. That concept is the goal plant community, and identifying it is more than selecting an archetype; it is the ability to see in your site the wild beating heart that wants to be expressed through planting. We begin with simple, universal landscapes like grasslands, woodlands, forests, or edge landscapes because they paint the big picture of what can be. These inspirations describe the basic elements, signature patterns, and general mood a site will convey. They are intentionally

This stylized dry meadow by Adam Woodruff features plants from all over the world, yet creates harmony because the plants come from related habitats. There is enough continuity of elements such as low grasses (genera *Sesleria*, *Eragrostis*, and *Molinia*) that even highly bred double plants like *Echinacea* 'Coconut Lime' feel natural.

flexible. The ultimate configuration can take almost any form. A grassland, for example, can be short or tall, wet or dry, loaded with colorful flowers or composed of a calm sea of grasses. The specific characteristics are determined by the site and your design objectives; but the idea that drives it is inspired by a more universal memory of nature.

The clarity of the initial inspiration is crucial. Part of the beauty of plants in the wild is how all the individual details—from the color of the soil to the textures of the plants—come together in one overall impression. Temptation to mix elements can be high. However, blending too many different landscape types comes with a high price: planting can feel cluttered and unfocused. Committing to a single archetype does not limit your design opportunities or the biodiversity of a planting. A grassland may have multiple expressions on a single site: a taller, more floriferous wet meadow at the bottom of a slope; a shorter, more uniform mix of grasses along a ridgeline; a subtle mixing of shrubs with grasses along a wooded edge. The more focused your inspiration is, the more powerful the final planting.

Of course, larger sites with diverse existing conditions may benefit from multiple landscapes goals. The Native Garden at the New York Botanical Garden is a four-acre site that was partially forested and partially open. The landscape architecture firm Oehme, van Sweden & Associates identified three central landscape goals: a forest, a forest edge, and a grassland. Much of the effort in developing the planting concept focused on how to seamlessly transition these different inspirations in a relatively small site. This transition had soft edges where certain highly adaptive species of ferns and sedges ran through both forest and grassland; it also had distinct edges where large hardscape features such as a boardwalk provided the physical barrier between landscapes. The success of the planting is in the seamless integration between dark forest and open meadow.

Balancing multiple goals on a single site is possible, but it can require more effort and artistry to create an authentic feeling. Some sites may actually want to express a transition between two archetypal landscapes. For instance, perhaps your site is located on the edge of a forest right before an open area. These types of settings are common around new developments shortly after existing trees have been cleared. You may want to select an archetype that connects both landscapes and creates a smooth transition. Selecting an open woodland or shrubland landscape goal may be a good decision in this specific case, allowing you to blend wooded and open areas in one continuous expression.

Species of *Helianthus, Lobelia, Bouteloua,* and *Sorghastrum* bloom at at the New York Botanical Garden's Native Plant Garden.

It helps to keep the initial inspiration for a planting as simple and pure as possible. Grassland, woodland, and forest (top to bottom) are three strong starting points that can be reinterpreted in endless ways.

The gradient of wet to dry is elegantly expressed in a stylized meadow at the New York Botanical Garden's Native Plant Garden.

These sketches by Marisa Scalera show the process of layering the meadow pictured on the left. In the upper part of the meadow, a matrix of low grasses forms the ground cover layer (below). Forbs are then added into the matrix (bottom), some dotted, some massed, and some drifted. The linear patterns of the drifts parallel the topography.

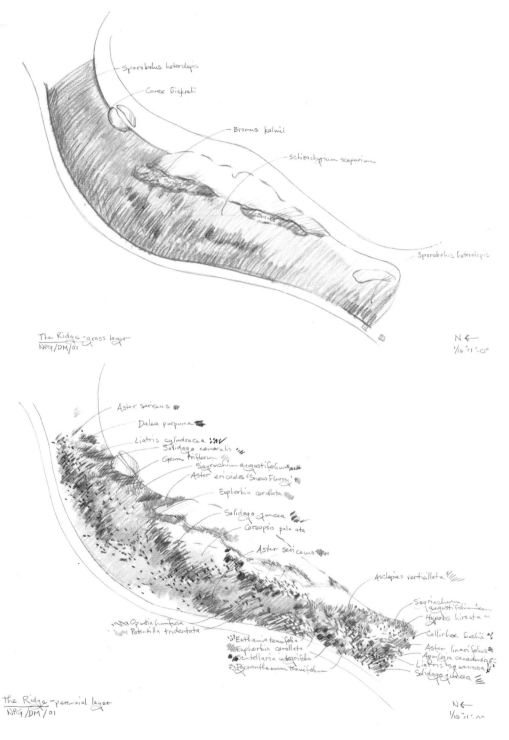

A street-side planting by Terry Guen Design Associates (left) uses formal rows of hornbeams and linear blocks of low perennials to create an orderly entry for a University of Chicago building. A rooftop planting designed by

HMWhite (right) embraces an exaggerated wildness to bring a feeling of prairie to the Manhattan skyline. Context is critical—these two urban plantings use similar plants but take two very different approaches to arrangement.

IDENTIFY HUMAN NEEDS AND CONTEXT

Planting exists to please people. This simple fact separates designed plant communities from their naturally occurring counterparts. Our method is inspired by the natural world, but that does not mean a naturalistic planting style is inherently better than other styles. A community-based approach to planting can accommodate any number of planting styles, from formal gardens to minimalistic modern plantings. As we will see later, much of the perceived style of a planting depends largely upon how its edges are framed and treated as well as the character of the hardscape. A low mixed planting surrounded by a symmetrical parterre of clipped boxwoods will read as formal garden; the same mix located in a raised Corten steel planter will appear modern. Successful plantings require translating natural patterns and process into human contexts.

Designed plantings encompass an entire gradient of different forms, from highly manipulated, ornamentally focused arrangements to ecological restorations entirely independent of human intervention. Understanding where in the gradient your planting falls is a critical part of developing the conceptual framework. First, how ornamental or functional should it be? Corporate clients and public parks may demand a certain level of tidiness and ornamental beauty from their landscapes. Regular maintenance can keep a mixed planting neat, though on many projects skilled gardening is often not guaranteed, requiring the designer to focus on species that look tidy during much of the year. Residential gardeners often expect a certain amount of color via flowers or foliage. As a result, grassland planting in a garden context would likely emphasize a heavier mix of flowering perennials than grasses. On the other hand, a storm water management planting located far away from buildings or roads might need to emphasize function more than ornament. In this case, a stable mix of vigorous grasses, sedges, and rushes could account for the bulk of that planting.

Understanding how formal or informal the planting should be is also an important contextual reference. More formal plantings may emphasize cleanly massed blocks of plants or large matrixes where a single species dominates for much of the year. These plantings can still be vertically layered with multiple species, but one or two species dominate visually. Informal plantings allow more visible mingling of species and self-seeding.

The final contextual gradient is that of urban to rural sites. Whether a planting happens in a city, a suburb, or the countryside will have a large effect on how we perceive it. More cosmopolitan mixes of plants from different parts of the world may look appropriate in an urban planting, but be jarring in a rural setting. A plum-colored cultivar of *Heuchera* mixed with ferns adds interest to a container, but in a more naturalistic setting, the colored foliage may feel dissonant with the natural hues.

The other piece of the human context is thinking about the long-term future of a planting. All landscapes change over time, so understanding how much natural evolution is desired is important. Will it have a climax state that is intended to persist, or will it dynamically evolve and transition into another landscape type? Answering this question early on can greatly influence the design process. You may consider a couple of options for managing change.

Colored foliage cultivars of genera like *Heuchera* give a more intentionally horticultural look to plantings.

Preserve the long-term design frame

Here, some species dynamics are allowed as long as they do not compromise the overall frame of the design. The proportion of filler and ground-covering species will likely shift, but the visually dominant species must remain where they were originally placed, since the legibility of the planting depends on them.

Allow succession to alter the goal landscape

In some cases, a gradual transition into other landscape types may be desirable. For example, a client may start out with a new homestead lacking any existing trees or shrubs. In this case, the planting may begin as a grassland and over time, grow into an open savanna or woodland landscape. Management practices have to take this into account and allow for the slow transition of a planting into a different archetype. The opposite may be the case after an unplanned disturbance of a site. For example, a dense forest planting suffers from windfall and the canopy opens up. In this case a planting transitions into an open woodland or forest edge archetype, at least until the canopy has closed again.

The essence of a wet meadow is competitive species such as upright forbs of genera like *Eutrochium*, *Hibiscus*, *Veronicastrum*, and *Helianthus*. Here, Oehme, van Sweden has exaggerated a naturally occurring wet meadow by repeating the most visually memorable species of that community in bold drifts.

SELECT, DISTILL, AND AMPLIFY CHARACTER-GIVING ELEMENTS

The irony of creating plantings that evince a sense of nature is that it requires a high degree of artifice. Literally transposing thirty square meters of a forest into an urban courtyard may not create the feeling of a forest at all, but rather just come across as a random assemblage of trees, shrubs, and ferns. Aptly conjuring the real deal requires distilling a forest into its most elemental forms: the repetition of tree trunks placed closely together and the textural mosaic of mosses and low woodland perennials. It is only when an archetypal landscape is distilled into its most basic forms that it recalls the reference community.

Exaggeration is at the heart of this process. Natural landscapes have impact because of their massive scale and the repetition of key patterns and processes over hundreds of acres. By comparison, our urban and suburban sites lack the size and context of their wild counterparts. In the wild, all of the details—sky, rock, soil, water, and plant—work together to create a rich sense of place. In contrast, buildings, roads, and cars often surround our designed landscapes. Our towns and cities are visually complex. In fact, our gardens are more likely to be surrounded by streetlights and power lines than waterfalls or boulder outcroppings. So in order to immerse a visitor in the feeling of a forest or grassland, we have to turn up the volume, creating designed plantings even more intense than their natural counterparts.

146

Oxeye daisies in a field of cool season grasses (left) is a recognized hallmark of British hay meadows. Sarah Price and Nigel Dunnett's design for the Olympic Park European Garden (right) amplified this mix with bolder patterns and a more floriferous *Leucanthemum* cultivar.

The key is to abstract the visual essence of these landscapes. In fact, abstraction is the heart of all art. Painters understand that rendering a landscape does not mean replicating every detail. Instead, abstracting is often more about removing irrelevant details and focusing only on those essential patterns or colors that give the landscape its power. All art is a process of selection, distillation, and amplification. These three steps form the basis of abstracting a wild plant community.

First, you select only those forms, patterns, or plants that reference the goal landscape. If you want to emulate a forest, a repetition of canopy trees is essential. No trees, no forest. So that becomes one element to select. This step defines the structural framework of a design. Think of it like the foundation and structural framework of a house. Next, you distill that element into its purest form. Perhaps you purge the site of mid-height shrubs in order for the tree trunks to read more strongly. This exaggerates the contrast of vertical and horizontal plane created by the trees and understory. Finally, you amplify the forms, patterns, or plants that connect you with your landscape goal. Perhaps you plant trees even closer together than they might occur in the wild, exaggerating the density and scale. Or you use trees with distinctive bark like birches, sycamores, or beeches to make the trunks stand out. Or you introduce a single thematic understory tree like a dogwood to add both a unifying element and a seasonal theme plant.

The more urban your site—that is, the more removed a site from a natural context—the greater the need to exaggerate and amplify the connection to the archetype.

< Plant communities may be mixed, but not all plants have equal weight in shaping one's visual impression of a place. Here wiregrass (*Aristida stricta*) dominates the ground plane, punctuated with islands of saw palmetto (*Serona repens*).

Do not be timid in creating design gestures that recall your archetype. Subtlety is not effective at this stage. Think about your design in the same way bonsai or penjing artists do. In a single tray, bonsai artists capture the essence of an entire forest. Only the essential forms of trees, rock, and moss are selected; all of the details convey a sense of place. The following two compositional strategies can help you amplify the character-giving elements of your planting.

Strategy 1: Use a high percentage of visual essence species

In established, naturally occurring plant communities, plant distributions are rarely equal. In fact, one of the distinctive qualities is the presence of a handful of dominant species within each habitat. In grassland communities, grasses often dominate; in forest communities, it may be a handful of characteristic trees, shrubs, and herbaceous plants. Ecologists often name plant communities after these dominant species (Loblolly pine/scrub oak woodland or heather moorlands). In garden terms, Richard Hansen referred to these as *Leitstauden* (leading perennials), or those that give us the visual impression of a planting and are instrumental in its ability to be understood.

Eastern skunk cabbage (*Symplocarpus foetidus*) is a visual essence species of wet, bottomland forests.

Because these species are so dominant, they often convey our mental image of those wild plant communities. An oak savanna without oaks or a heather moorland without heather loses its meaning and sense of place. To determine these plants, visit nearby examples of plant communities of your selected landscape archetype and look at the characteristic species. This will help you refine your archetype with a regionally appropriate palette. For example, if you selected a forest as your archetypal landscape goal, your interpretation might be an open oak-hickory forest because this type of forest frequently occurs in your region. Oak and hickory species should then be the visually dominant species in your design. Visual essence species occur in all layers of a planting. An open woodland in Virginia would likely feature the tree *Juniperus virginiana*; a desert grassland in Arizona would use the annual *Eschscholzia californica*. The corollary is to avoid using species from archetypes different than your goal landscape. If the goal is an open woodland, for instance, remove pure meadow species, since they send the wrong message. Species such as *Ruellia humilis* or *Eragrostis spectabilis* look authentic in a sunny meadow, but they have the wrong leaf color and texture for shady sites and dilute the end result.

SPECIES THAT CONJURE ARCHETYPAL LANDSCAPES

All of these species have powerful associations with their unique habitats of origin. They are essential design tools and suggest connections with larger landscapes.

⩗ *Echinacea simulata* powerfully evokes wild prairies.

⌃ *Asclepias tuberosa* is at home in xeric grasslands.

⩗ *Asclepias incarnata* reminds us of colorful wet meadows.

⌃ *Sisyrinchium angustifolium* 'Lucerne' is reminiscent of floodplain landscapes.

⩗ *Salvia nemorosa* 'Caradonna', *Nepeta*, and *Nassella tenuissima* originate from dry grasslands.

⌃ *Eurybia divaricata* and *Thelypteris decursive-pinnata* evoke forest landscapes.

Adam Woodruff's design for the Jones Road garden uses solid blocks of plants like *Calamagrostis* ×*acutiflora* 'Karl Foerster' in the background to set off the more intricate planting in the foreground, layering the space and making it feel deeper.

Strategy 2: Make plant patterns visible

Established plant communities often have visually stunning patterns. Sometimes patches of one species aggregate into little archipelagos in a sea of a more dominant species. Other times, dense blocks of clonal spreading species form a quilted-looking landscape. Or evenly dispersed arrangements of accent species create an intricate textural mat. These patterns are important, not only for their beauty, but also for the clues they reveal about how the plants compete and coexist.

Designers can stylize natural distribution patterns in several ways. The simplest way is to create tighter, denser, and larger versions of the patterns than those in the wild. Example: If wild aster forms loose drifts through a field of grasses, then perhaps these drifts are represented by a thicker mass of asters in a designed planting. Or if a *Liatris spicata* dots itself singly through a prairie, perhaps a designed planting would use clusters of five or seven *L. spicata* plants to create a more robust, readable version than what happens in the wild. To achieve the same effect in a smaller scale garden, we have to use more of the same species and at higher density.

Massing species together is one of the most important tools a designer has to express natural patterns in more artistic, coherent ways. Massing should be considered based upon the species and how it naturally is arranged in the wild. Sociability—that is, how far plants within the same population grow from each other—offers a good model for distinguishing which plants should be massed versus which should be placed individually. The German researchers Hermann Müssel, Rosemarie Weisse, Friedrich Stahl, and Richard Hansen organized plants into five categories, with solitary specimens (1) on one end of the spectrum and clonal-spreading ground covers (5) on the other. Plants on the lower end of the sociability scale (1 and 2) are generally tall and visually dominant, and should be arranged individually or in small clumps of three to ten. *Asclepias tuberosa* or a species of *Echinacea*, for example, are almost always found scattered individually in the wild. On the other hand, plants on the higher end of the sociability scale (3 to 5) are excellent ground covers that should be arranged in masses of ten to twenty or more. The perennial *Tiarella cordifolia* or the low woody *Vaccinium angustifolium* are examples of plants with high sociability. Species high on the sociability scale (4 to 5) often have shapes and behaviors that make them good ground covers, allowing them to be used in large quantities under taller perennials. Levels of sociability help us understand what species might work as structural plants or seasonal theme plants.

DEVELOP A DESIGN STATEMENT

Once you are clear about your goal landscape and elements to amplify, the next step is to define how that landscape will be applied on your site. It is helpful to write down a few of the actions needed to bring your long-term landscape goal to life, similar to a corporate strategic statement. This glorified to-do list will help create the larger spatial framework of a planting. It describes the design responses that will guide the application of your archetype to your site. Make these statements short, direct, and action oriented. The goal is to crystallize key activities that will ensure the integrity of your concept. Such statements become the filter for all future design questions; they will keep you on track and prevent major design mistakes.

Do not focus on plant selection yet. What matters here is understanding the patterns of open areas and closed canopy, isolating the different compositional layers, and describing the character of the final planting. Removals should also be identified—replacing a thicket of tall invasive shrubs with a mosaic of low perennials, or eliminating trees to expand a meadow. Ultimately, design statements always refer back to the goal landscape, focusing on the actions that bring an archetype to life.

It is important to remember that planting design is the cultivation of an idea as much as it is a physical site. The time spent selecting a landscape goal and developing it conceptually can save significant time on the back end of a design. Too many planting designs leap from site analysis directly to plant selection. Charting the big moves not only sharpens the idea, but provides criteria for plant selection that will save time and effort. The chart on page 154 gives three examples of how design response statements might emerge from understanding your site and goals.

LEVELS OF SOCIABILITY

Plants can be characterized by their level of sociability. Illustration after Hansen and Stahl, 1997.

LEVEL 1	LEVEL 2	LEVEL 3	LEVEL 4	LEVEL 5
INDIVIDUAL PLANTS OR SMALL GROUPS	SMALL GROUPS OF 3 TO 10 PLANTS	LARGER GROUPS OF 10 TO 20 PLANTS	EXPANSIVE GROUPS	PRIMARILY LARGE AREAS

Aruncus dioicus	*Caltha palustris*	*Achillea millefolium*	*Allium cernuum*	*Carex pensylvanica*
Eryngium yuccifolium	*Coreopsis verticillata*	*Aquilegia canadensis*	*Carex morrowii*	*Conoclinium coelestinum*
Eutrochium fistulosum	*Deschampsia cespitosa*	*Asarum canadense*	*Carex plantaginea*	*Geum fragarioides*
Heliopsis helianthoides	*Echinacea purpurea*	*Bouteloua curtipendula*	*Chrysogonum virginianum*	*Hypericum calycinum*
Panicum virgatum	*Liatris spicata*	*Geranium maculatum*	*Erigeron pulchellus* var. *pulchellus*	*Packera aurea*
Solidago caesia	*Monarda fistulosa*	*Heuchera longiflora*	*Mertensia virginica*	*Sedum spurium*
Sporobolus wrightii	*Pycnanthemum flexuosum*	*Monarda didyma*	*Onoclea sensibilis*	*Stachys byzantina*
Vernonia noveboracensis	*Symphyotrichum laeve*	*Rudbeckia fulgida*		*Tiarella cordifolia*

Sporobolus wrightii

Pycnanthemum flexuosum

Bouteloua curtipendula

Mertensia virginica

Carex pensylvanica

CONCEPTUAL FRAMEWORK FOR A DESIGNED PLANT COMMUNITY

1. EXISTING SITE	2. GOAL LANDSCAPE (ARCHETYPE)	3. ELEMENTS TO AMPLIFY	4. DESIGN RESPONSES (ACTIONS)	5. ORDERLY FRAMES (HUMAN CONTEXT)
Steep slope with mixed deciduous trees; lots of invasive shrubs and vines on ground	Forest	Repetition of mixed hardwood trees Diverse, colorful mosaic of perennials on woodland floor	Remove invasive shrubs; replant with native shrubs along the perimeter of the site to screen site. Repeat a mix of deciduous trees with strong fall color. Establish an understory tree layer focusing on spring bloom. Layer the ground plane with a mix of sedges, ferns, and forbs.	Informal hedge along road Stone step path down slope Dominant theme plants in various seasons
Suburban lawn with scattered trees adjacent to a busy road	Woodland Edge	Layered planting tiered in height Richly interlocking compositions of woody and herbaceous plants	Plant a spine of taller shrubs to screen the road in the background. Layer a mix of tall perennials and ferns between shrubs in the mid-ground. Create a colorful foreground of low perennials.	Emphasize screening plants with seasonal themes (flowers, fall color) Smooth transition of height from tall shrubs to tidy, low perennials along path
Large mown lawn in an office park, small trees and shrubs appear along edges like a shrubland succession	Grassland	Low, even-height meadow Drifts of blooming perennials in different seasons	Remove woody species. Establish a matrix of low grasses. Create a succession of color with a series of seasonally dominant perennials.	Lawn verge along edge of meadow Neat massings of low grasses along edge of planting

Calamintha nepetoides forms a clean edge against a lawn in author Thomas Rainer's garden.

PHYSICAL FRAMEWORK: CREATE "ORDERLY FRAMES" FOR MIXED PLANTING

Once your conceptual framework is clear, the next step is to design the physical framework of the planting. Here it's good to apply the concept of "orderly frames" mentioned earlier. We want to focus on a range of techniques that can help mixed plantings relate to their human context.

The shape of planting beds

A designed plant community can be almost any size or shape. In a small residential garden, the planting may be no larger than a single bed. In smaller or more urban sites, giving planting beds a strong geometric shape is one way of signaling its intentionality. The simpler the shape, the better, particularly when the planting itself is rather complex. For an urban site adjacent to buildings or other structures, a simple rectilinear bed creates a clean frame that relates to the context. In fact, thinking of planting beds—clearly defined, rectilinear beds with an interlocking mosaic of plants—as a beautiful carpet is a helpful way to understand how these plantings can be placed in urban contexts. In a suburban setting with a larger-sized yard, broad curvilinear beds may be more appropriate. If this geometry feels more appropriate for your site, try to use simple, large radii curves rather than tight, wavy bed lines. Bed lines always look best when they are big gestures. A single curve or a broad, gently curving "S" shape can add plenty of mystery and complexity to a site. Overly complex curves create planting beds that look forced and trite; more like a miniature golf course than a naturalistic landscape.

Restrain the height of plantings

Height control is one of the most effective ways plantings can fit in urban contexts. Environmental psychology has long documented that people like spaces best when they have long views over them. Thus, when planting exceeds waist or chest height, it can appear overwhelming. Obviously, there are important exceptions, such as the need to screen unsightly views, but by and large, a clearly defined bed that people can look over is much more acceptable. While the majority of planting is best kept below waist height (eighteen to thirty inches tall), occasional accent species can go much higher, particularly when these species have taller leafless stems that people can see through, such as *Rudbeckia maxima*, or species of *Molinia* or *Sorghastrum*.

Create a frame around plantings

A literal frame around the planting itself, in the form of some clearly defined edge, is one of the most effective ways to make a wild planting appear intentional and fitted to a site. This can take any number of forms. Lawn, for example, is a classic foil to planting

A matrix of low grasses such as *Bouteloua curtipendula*, *Eragrostis spectabilis*, and *Schizachyrium scoparium* blend with species of *Perovskia* and *Eupatorium*.

beds. A simple lawn verge around a meadow style planting adds a clean border that signals care. In American gardens, where front lawns are such a dominant element of the vernacular, designed plant communities may be placed next to lawns—not replacing them entirely. In this way, lawn and planting beds can be somewhat symbiotic, each improving the visual quality of the other.

Hardscape and other architectural garden features such as walls, hedges, and fences also create attractive frames around planting. Traditional perennial borders have long used cobble edging, boxwood parterres, and yew hedges as frames for more complex, layered herbaceous plantings. These framing strategies are particularly useful in smaller, more formal settings such as courtyards and urban gardens. Finally, paths can serve the double function of providing access through plantings—allowing for easier maintenance—as well as a sharply defined edge. For plantings in rural settings where hardscape or clipped hedges are not practical, paths are especially important for defining edges. In an open setting, a simple mown path is an effective way of separating designed

Strips of low, mixed meadow planting help relate this design
by Terry Guen Design Associates to the urban context.

from wild planting. In a woodland setting, mulch paths can help to define zones where
planting is most intense.

A frame can also be formed by plants themselves. It could be a band of shorter
species around an urban meadow, such as *Sesleria autumnalis*, *Calamintha nepetoides*,
or *Amsonia tabernaemontana* around a grassland planting. Or the outer five feet of a tall
meadow can be mowed in early summer to keep vegetation on the outer edges short and
neat. This technique is used by American meadow expert Larry Weaner along pathways
through taller meadow communities. Vegetative frames are just as effective as architectural ones, and often less clumsy.

All of these strategies frame and package new forms of plantings into what Nassauer calls the "vernacular language of landscape." Framing planting helps define context. Ultimately, however, it is the planting itself that must be functional and beautiful
in order for it to be successful. The next chapter will examine how plants are layered and
combined to create enduring plantings that harken to nature.

⩓ A rail fence helps frame the planting and relate it to its rural context.

⋀ Low grasses such as *Schizachyrium scoparium* frame the edge of this storm water planting. Taller perennials like *Veronicastrum virginicum* are kept in the center of the bed.

⩓ Low plants with a consistently neat appearance were planted along the edge of a sidewalk to frame a seeded mixed meadow. Plants like *Sesleria autumnalis* and *Amsonia* 'Blue Ice' are good choices as planted frames.

⋀ A bench and low grasses in the planting itself create a frame.

The relationship of plants to other plants is what defines a community. Here we put aside traditional landscape techniques designed to avoid plant interaction—heavy use of mulch, pruning, and wide plant spacing—in favor of a tightly knitted mosaic of different plants inhabiting different niches. Only with a clearly defined landscape goal and well-designed framework will a planting shine and reveal its emotional power.

RELATING PLANTS TO OTHER PLANTS

Plant selection flows out of your design framework. Think about your design as a series of empty frames that will be filled with various layers of plants. Instead of jumping directly to plant lists for dry soil or full shade site conditions, we will first determine which plant types are suitable to fill specific parts of our empty design and frame with vegetation.

Our approach to selecting plants focuses on layering a landscape vertically. This differs from the conventional approach of planting in monocultural blocks—one plant in one place. We prefer to think of plants in various layers on top of each other, both in space and time. This layered approach allows you to achieve a fantastic density and diversity of plants, but in a clearly intentional plan. What defines the various layers is both the behavior of a plant—that is, how it grows and competes with other plants—and the design framework.

PLANT SELECTION TOOLS: DIFFERENT PLANT STRATEGY SYSTEMS

Before describing the various layers of plants in our system, we would like to briefly look at a few international examples of plant strategy systems. For generations, plant professionals and designers have tried to classify plants with the goal of developing general recipes for good planting design. Many different approaches exist, ranging from purely empirical to science-based plant classification. Even though none of these has been able to create a perfect method for planting design, simplifying and combining the best aspects of each can help tremendously.

Each of the three systems we present here provides tools for the designer to translate the massive quantity of species and cultivars presently available on the market into usable elements of planting design.

Thought leaders of the plant habitat–focused system

In their revolutionary book *Perennials and their Garden Habitats* (1979), Richard Hansen and Friedrich Stahl put forth that if species are planted in conditions that are similar to their wild habitats, they would live longer, be more resilient, and be easier to manage. The idea is that if you combine plants from similar habitats—a sage from the Mediterranean, some annuals from the California chaparral, and grasses from a Eurasian scrubland—then you could create a stable, but entirely novel, plant community. Hansen suggested that carefully selected plant palettes would form a *lebendige Bodendecke*

In the wild, plant populations are often layered one on top of another, such as mayapples (*Podophyllum peltatum*) growing through a matrix of Pennsylvania sedge (*Carex pensylvanica*).

(living ground cover) that is mostly self-regulating. Extensive long-term trials of plants in Weihenstephan, Germany, proved some success of this method.

Hansen's habitat system provides essential insights for creating communities. However, all of the plant lists offered in the book were based on species available in Europe at the time. Translating his lists into regionally appropriate examples for other parts of the world takes a high degree of plant knowledge. Moreover, many internationally renowned designers, like Piet Oudolf and Petra Pelz, have created successful plant communities with plants from different habitats. This strategy does not really explain how plants from different habitats—meadows and forests, for example—can still form effective designs. But a different method evolved in European ecology than can solve some of the limitations of this habitat approach: Grime's universal adaptive strategy theory (UAST).

From above, this traditional mass of *Amsonia hubrichtii* looks full, but when viewed from the side, there are many empty spaces waiting to be occupied by other plants.

John Philip Grime: plant survival strategies

The British ecologist John Philip Grime's strategy explains the behavior of plants within natural settings. His research focuses on the limited resources species face in their habitats and how plants adapt to these. He described three limitations plants face in the wild: strong competition from other species of the same community, stress conditions like drought or shade, and high levels of external disturbance (also called ruderal influences) like fire or herbivory (plants being eaten). All three forces cause different reactions and survival strategies. A plant may decide to throw most of its resources toward growth, maintenance, or regeneration, depending on what factors limit its development. Grime found that there is always a three-way trade-off between the allocation of resources, and his C-S-R theory divides plants into three categories depending on how well they respond to competition, stress, and ruderal influence.

Richard Hansen's ideas continue to be tested and further developed at Hermannshof, in Weinheim, Germany, one of the most influential test and trial gardens in the world. The garden combines plants from similar habitats around the globe, such as this one inspired by the Eurasian steppe.

Competitors include long-lived, clonal-spreading plants like species of *Eutrochium* that hold their ground.

Stress tolerators include plants that can endure drought and infertile soils, such as species of *Artemisia* and *Ajania*.

Ruderals include plants like annuals and many garden weeds that quickly colonize disturbed soils, but are poor competitors.

C: Competitor. Plants of this category thrive in habitats of low stress and disturbance. Perfect growing conditions attract many species, which is why competition in such areas is high. In order to survive, plants of this category are very good at outcompeting other plants. They efficiently use available resources and evolve highly adaptive strategies, such as rapid growth and high productivity. This category includes popular prairie plants and species of mesic meadows.

S: Stress tolerator. In order to survive in areas of high stress intensity and low levels of disturbance, plants allocate their resources to maintaining biomass. Characteristic responsive strategies include slow growth and physiological variability. Successful stress tolerators keep their leaves for a long time and show few seasonal changes in order to retain nutrients. This category includes green roof species and spring ephemerals.

R: Ruderal. Ruderals are at home in areas of high levels of disturbance and low stress intensity. Successful survival requires fast growth between events of disturbance. Species of this category are able to complete an entire life cycle in a very short time and they focus their resources on regeneration, often producing massive amounts of seeds. Some of our most adored annuals are ruderal species, however, dreaded garden weeds often fall into this category as well. Ruderals frequently colonize recently disturbed floodplains or newly tilled garden beds.

Grime's C-S-R strategy can be a powerful tool in creating designed plant communities for extreme sites. For example, an urban street planting suffers from consistent disturbance caused by foot traffic, dogs, and street cleaning equipment. Therefore, a combination of R or ruderal choices will most likely lead to a balanced and long-lived design that can repair itself after disturbance. The drawback of Grime's model is that very few plants fall purely into these three categories; many have traits of all three. It is useful as a conceptual model, but it offers little actual guidance for combining plants.

Norbert Kühn: plant strategy type model

German professor Norbert Kühn recognized the strengths and limitations of Hansen's and Grime's models. By evolving both models to include a plant's reaction to site conditions, its morphology, its propagation and spreading behavior, and its temporal niche, Kühn transformed and blended them into a very promising tool for creating designed plant communities. For the first time, a plant's adaptive strategy was included in the model. A plant may endure stress or it may avoid it altogether. If all possible combinations of site conditions and plant adaptations are laid out, the model becomes very complex and loses its clarity and usefulness for planting design. Therefore, Kühn focused on the most relevant scenarios typical for garden settings, and narrowed adaptive strategies down to eight main categories.

In his 2011 book, *Neue Staudenverwendung*, Kühn breaks each type into several subcategories, allowing for more accurate classification and better guidance when using this tool to choose species for the parts of a designed plant community. For example, a new planting that will receive little maintenance might use tall perennials of Type 4 as a design layer, and ground-covering plants of Type 5 underneath, resulting in a planting that needs little management. Elements of these plant classification systems will help you select and combine compatible species in your planting designs.

THE VERTICAL LAYERS

Taken together, the three systems described offer designers a range of tools to select and combine plants. All are plant-centric models; they put plants and their relationships with the environment first. Yet in terms of planting design, these systems do have weak points. For instance, all require a high level of knowledge to translate into plant selection. Richard Hansen's model is the most focused on the designer, yet his model relies heavily on regionally specific plant lists. All three systems also emphasize the functional arrangement of plants, providing little guidance about their aesthetic arrangement. Finally, all of the models are conceptual, leaving much of the application to the designer to interpret.

Our goal is to offer a simplified approach that extracts the most relevant aspects of each of these systems. We start by thinking of plants not as types or categories, but instead as a series of layers that are sequentially added to the site. The layers are arranged vertically, like the stories of a building, each separate and distinct until ready to be combined on the site. The process of plant selection starts with the tallest, most visually distinct layers, and progressively moves down to the lower, more functional layers. The aforementioned plant strategy systems will then help fill each layer densely with plants of appropriate behavior, habitat association, and survival strategy. However, before describing the specific layers, it is important to understand the two categories of layers: design and functional.

Design layers include plants that are easily visible from eye level, such as this common boneset (*Eupatorium perfoliatum*).

Design layers

Design layers describe the tallest, most visually dominant species within a community. These are the plants which form your impression of the landscape. They draw your attention with their distinct architecture, tall height, and bold colors and textures. The design layer typically includes trees, dominant shrubs, and tall perennials and grasses. They include distinctly structural elements like evergreen trees or tall upright grasses, and also carry big seasonal moments in a planting—such as the flush of color from asters in a fall meadow, or a mass of rhododendrons blooming next to a mountain stream. While plants in this layer are the most visually prominent, they are

CATEGORY	EXAMPLES	DESCRIPTION
TYPE 1 Strategy of conservative growth	*Lavandula, Santolina, Phlox subulata*	Plants of this category grow slowly and consistently. This group includes chamaephytes of low height and often creeping growth habit. They can be found on extreme sites with very limited resources, such as rock outcrops, xeric meadows, and alpine sites. Competition from other species is very limited or absent. If site conditions are improved by gardeners, these plants often react with shorter life expectancy.
TYPE 2 Strategy of moderate stress adaptation	*Aquilegia, Hosta, Salvia nemorosa*	Site conditions may limit a species' development through lack of sunlight, water, and nutrients. Species of this group compete within stressful sites, however if planted in ideal conditions, they may lose their unique morphological stress adaptations, such as silver leaves, large leaves, or longevity.
TYPE 3 Strategy of stress avoidance	Early-flowering forest species and spring geophytes. *Helleborus, Crocus, Allium christophii*	This category includes spring geophytes which avoid stress altogether by completing their life cycles very quickly during times of optimal growing conditions. They avoid unfavorable conditions and stress by going dormant. Gardeners and designers use this group of plants to extend the garden season.
TYPE 4 Strategy of area occupation	Tall perennials. *Rudbeckia fulgida, Phlox paniculata, Helianthus microcephalus*	Perennials of this group derive from sites with excellent growing conditions, such as floodplain meadows and tall grass prairies. Competition within these sites is high and survival depends on the ability to stand one's ground. Designers use plants of this category as structural frames for perennial plantings.
TYPE 5 Strategy of area coverage	*Geranium sanguineum, Pachysandra, Ceratostigma*	Kühn includes carpet-forming ground covers of low height in this category. They are often found in forest edge habitats and their survival strategy is dense cover of all available habitat space. Planting designers use these either in masses or under species of strategy Type 4, like a layer of green and living mulch.
TYPE 6 Strategy of area expansion	*Ajuga, Lamiastrum, Solidago rugosa, Lysimachia clethroides*, and *Eutrochium maculatum*	This category includes species of aggressive, clonal-spreading behavior. This strategy is a response to highly dynamic site conditions; it allows a species to quickly cover new ground. Includes both low ground covers and tall clonal perennials.
TYPE 7 Strategy of niche occupation	Meadow plants like *Salvia pratensis, Narcissus poeticus*, and *Colchicum autumnale*	This adaptation strategy is especially successful in open habitats, such as managed meadows and hayfields. After the ground warms in spring, species of this type develop rapidly and surprise with vibrant summer colors. If cut back after the fast completion of their first growth cycle, they often flower again in late summer or fall.
TYPE 8 Strategy of gap occupation	Ruderals like *Erigeron annuus, Digitalis purpurea*, and *Gaura lindheimeri*	Plants of this type are naturally short-lived and produce large amounts of seed. They are highly dynamic and acclimated to sites of frequent mechanical disturbance, such as coastal, urban, or floodplain habitats. Species tolerate little competition and they will disappear if disturbance stops. If managed correctly, their dynamic nature can create refreshing surprises within designed plant communities.

not always the highest percentage of a planting. The reason we call it the design layer is because its goal is to create visually pleasing horticultural effects. While it certainly has ecological function, the focus from the design point of view is aesthetic arrangement.

Functional layers include low, ground-holding plants that grow at the base of taller, design-layer plants. In this case, *Carex pensylvanica* covers the base of *Asclepias purpurascens*.

Functional layers

The functional layer describes the mix of low, ground-covering species. Unlike the more visible design layer, almost no one sees it. Its purpose is to hold the ground and fill any gaps to prevent weed invasion. This creates conditions for some stability, so that longer-lived species in the upper layers can get established. It is composed of low, hummocky plants, many of which are somewhat shade tolerant. Ground layer plants are really nook and cranny plants—that is why they are functional. They have the unique ability to squeeze in between the space left over between the dominant plants. They are often self-seeding ruderal species, low-spreading plants that move through the ground like vines, or legumes that fix nitrogen in the top soil horizon. They have the ability to sequester carbon, control erosion, build soil, and provide nectar sources for pollinators.

LEGIBILITY IN THE DESIGN LAYER, DIVERSITY IN THE FUNCTIONAL LAYER

Understanding the distinction between design and functional layers is crucial to balancing beauty with function. Aesthetically pleasing design can be highly complex and diverse. Designers have great flexibility to create patterns or dramatic seasonal moments with plants in the design layer; all the while, the less visible plants in the ground layer provide diversity and ecological function. Our mantra for planting design is to create legibility in the design layers and diversity in the functional layers.

The German landscape architect Heiner Luz developed a design strategy that creates legibility on a large scale by using a large number of seasonal theme plants in highly attractive masses. His concept *Prinzip der Aspektbildner* (Principle of Aspect-Forming Plants) involves selecting a series of three to six seasonally spectacular theme plants that flush in a series of succession. In his design for Ziegeleipark in Heilbronn, Germany, Luz layers salvia and iris, for an explosion of color early in the season; later, species of *Aster* and *Echinops* create another spectacle. Underneath these theme plants, a handful

Heiner Luz's Ziegeleipark in Heilbronn, Germany, shows the tremendous ornamental potential of designed plant communities. Luz has layered this planting so that one wave of color follows the next in short succession: (top to bottom) *Salvia nemorosa* 'Caradonna', *Dianthus carthusianorum*, *Iris germanica* cultivars, and other irises bloom in late spring. Later in the season, *Aster × frikartii* 'Wunder von Stäfa' and *Stipa pennata* create additional seasonal themes.

	LAYERS	PERCENT	EXAMPLES	DESCRIPTION
DESIGN LAYERS	Structural / framework plants	10–15 percent	*Andropogon gerardii, Asclepias incarnata, Carnegiea gigantea, Cercis, Juniperus virginiana, Lindera, Quercus, Sorghastrum, Veronicastrum*	Large plants that form the visual structure of the planting. This includes trees, shrubs, upright grasses and perennials, and large-leaved perennials. Plants in this layer have distinct forms (silhouettes) and are long-lived. Plants tend to be competitors or stress-tolerators.
	Seasonal theme plants	25–40 percent	*Amsonia, Aster, Hemerocallis, Iris, Mertensia, Rhododendron, Rudbeckia, Salvia, Solidago*	Mid-height plants that become visually dominant during a season because of flower color or texture. When not in bloom, plants in this layer become green supporting companions to the structural plants. Long to medium lifespans. Plants tend to grow in masses or drifts. Competitors, stress tolerators, and ruderals can fit in this category.
FUNCTIONAL LAYERS	Ground cover plants	Approximately 50 percent	*Carex, Geranium,* geophytes and ephemerals like *Narcissus* and *Crocus, Heuchera, Packera, Tiarella, Waldsteinia*	Low, shade-tolerant species used to cover the ground between other species. Functions as ground cover, erosion control, nectar source. Plants tend to be rhizomatous. Stress-tolerators.
	Filler plants	5–10 percent	Annual *Erigeron, Aquilegia, Coreopsis, Eschscholzia, Gaura, Lobelia cardinalis, Stylophorum*	Ruderal and short-lived species that temporarily fill gaps and add short seasonal display. Plants grow quickly, but do not tolerate competition. Annuals, biennials, and short-lived perennials.

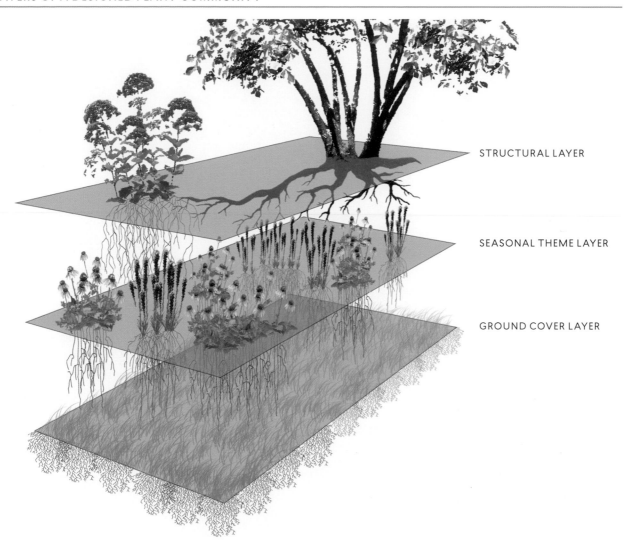

STRUCTURAL LAYER

SEASONAL THEME LAYER

GROUND COVER LAYER

of lower companion plants like *Dianthus carthusianorum* and *Stipa pennata* cover the ground, providing more subtle intricacy of color and texture.

Perhaps more than any other planting designer of our time, Luz is a master of balancing design and functional layers. Luz uses the motto "uniformity in the large scale, and variety in the small scale" to describe his approach to creating dramatic, large-scale effects with the design layer, while getting ecologically important diversity in the smaller scale, ground-covering layer. It is a perfect example of a diverse, layered approach to planting that refuses to sacrifice design clarity or bold ornamental quality.

Once the distinction between design and functional layers is clear, we can move on to the various layers of a plant community. In the design layers, we identify visually dominant species. In the functional layer, it is ground-covering species that make a planting a true community.

Long-flowering structural perennials such as *Lythrum virgatum*, *Veronicastrum* 'Fascination', and *Agastache foeniculum* make a striking contrast against the softer, filler forms of the genus *Deschampsia*.

LAYER 1: STRUCTURAL AND FRAMEWORK PLANTS

Structural plants form the backbone of the planting. They are the visual essence species of a community, and include large plants such as trees, dominant shrubs, and even tall perennials and grasses. In forests and woodlands, the structural plants are most often the trees themselves, forming the living architecture of canopy and evergreen walls. Shrubs also are important structural elements, particularly those with distinctive forms. In grassland communities, structural plants are often solitary upright grasses like *Sorghastrum nutans*, *Andropogon geradii*, or even *Miscanthus sinensis*. They also include

174

tall perennials like *Aesclepias incarnata*, *Veronicastrum virginicum*, or *Silphium terebinthinaceum*. Tall perennials with mostly leafless upper stems make great candidates for structural plants, as they allow light to penetrate to the lower levels of a designed plant community.

What distinguishes this layer is an emphasis on plant form. Structural plants tend to have distinctive shapes, in contrast to the more amorphous forms of filler plants. If you are not sure about a plant, think about its silhouette. If it is distinctive—the upright spire of a cedar, the spikey globe of a thistle, or the distinctive candelabra of a yucca bloom—it is likely structural. Piet Oudolf's brilliant work has long contrasted highly structural plants against more lacy filler forms, to great effect. The advantage of structural plants is their long season of interest, a sort of anchor for the eye to rest on in a planting. Even herbaceous plants can provide structure in winter, through the dried forms of tall grasses or the dark seed heads of forbs.

The structural layer creates the image of the community. At the canopy and shrub level, structural plants are often numerically dominant; but at the herbaceous layer, structural plants usually take a much lower percentage of the overall planting, often less than 15 percent of the total. For this reason, herbaceous structural plants rarely form seasonal themes because there are so few of them. Structural trees can indeed create seasonal themes. Think of spring-blooming redbuds or the fall foliage of maples.

Stability and reliability are key characteristics of the structural layer. These species ensure that the important framework of the planting will endure. For this reason, choose species that have the following characteristics.

Longevity

Structural frame species must be long-lived. Hybrids of *Echinacea* and *Coreopsis*, for example, are not good choices for this layer. Both generally only get a few years old and would require regular replanting in order to keep the design structure alive.

Little information about plant longevity is available to designers, and most available data is purely anecdotal or based on observation in home gardens. If you do not know the life expectancy of a plant, connect with a nursery or other local plant professional. Finally, species longevity very much depends on a site's growing conditions. Characteristically long-lived species like *Liatris spicata* and *Eragrostis spectabilis* can be very short-lived in heavy and nutrient-rich soils.

Long-lasting species are often slow to establish. This is an important consideration when specifying the size of a plant. For example, *Baptisia australis* can take up to three years before it reaches mature width and height. This species channels most of its energy into underground storage organs first; only when its reserves are fully formed will it put on significant aboveground growth and flowers. Once its underground backup system is in place, this species is highly resilient. It will readily grow back after disturbance such as fire or mowing. As a result, consider plants with larger, more established root systems for plants in this layer. Seeding or using plugs will likely take a few years to achieve mature plants.

⋏⋏ *Panicum virgatum* 'Northwind'

⋏ *Asclepias incarnata*

⋏ *Liatris spicata*

⋏⋏ *Rudbeckia laciniata* 'Autumn Sun'

⋏ *Thalictrum rocheburianum*

⋏ *Vernonia noveboracensis*

⋏⋏ *Vernonia glauca*

⋏ *Veronicastrum virginicum*

⋏ *Sporobolus wrightii* 'Windbreaker'

Clump forming

Use clump-forming or at least slowly spreading species in the structural layer. Avoid rapidly spreading plants like *Physostegia virginiana* or *Pycnanthemum muticum*, which will multiply over time and erode the clarity of this layer. Clump-forming grasses like species of *Panicum*, *Andropogon*, and *Sorghastrum*, in addition to well-behaved forbs like *Vernonia noveboracensis*, are ideal herbaceous structural plants.

Year-round structure

Choose plants with reliable aboveground structure. The most successful structural species withstand snow and ice storms in the winter as well as wind and rain storms in the summer. Tall perennials with mostly leafless upper stems make great candidates. The lack of leaves on the upper parts of the stems provides little surface for rain or snow to weight down. Grasses like *Sorghastrum nutans*, species of *Molina*, and *Stipa gigantea* persist even in bad weather. Perennials like *Phlomis tuberosa*, *Aster tataricus*, and *Rudbeckia maxima* can hold their stems even in heavy snows.

LAYER 2: SEASONAL THEME PLANTS

The next layer, seasonal theme plants, is made up of companions to structural plants. This layer focuses on plants that visually dominate the planting for a period of time during the year. This might be demonstrated by the dramatic seasonal flowering of iris or aster in a meadow, or the bold texture of *Podophyllum peltatum* on a forest floor. Seasonal theme plants take the visual lead at certain times of year, and disappear back into the green background after the show. This does not, however, mean that they fade away and leave gaps in the planting. The opposite is true, in fact: they continue to cover soil and act as companions to the structural layer. They are used in larger quantities in order to create stunning effects of color and texture. In sunny open sites, perennials of genera like *Salvia nemorosa* and species of *Solidago*, *Calamintha*, and *Hemerocallis* all are solid companion plants with strong seasonal displays. In forests, this layer may be composed of textural plants like the fern *Dryopteris erythrosora* 'Brilliance' or flowering perennials of genera like *Aruncus* or *Actaea*.

Since plants of this category appear in larger quantities (25 to 40 percent of planting), exact placement of each individual is less critical than for structural species. Individual plants melt into larger strokes of color and texture. Theme plants can be shorter-lived and more dynamic than plants that build permanent frames, because the objective is keeping their population alive and not keeping them growing in exactly the same spot. However, if theme plants disappear or multiply too much, the entire planting suffers. Therefore, species of medium longevity and vigor are well suited to fill this part of a designed plant community.

In contrast to the sharp silhouettes of the structural layer, seasonal theme plants are often amorphous in shape. Their role as filler plants is to spill over and around structural plants, occupying the gaps between architectural forms. These companion plants

⚔ *Chrysopsis villosa*

⚔ *Monarda bradburiana*

∧ *Achillea* 'Strawberry Seduction'

⚔ *Ratibida columnifera* 'Red Midget'

⚔ *Amsonia* 'Blue Ice'

∧ *Coreopsis* 'Red Satin'

⚔ *Caryopteris* ×*clandonensis* 'Inoveris'

⚔ *Symphyotrichum oblongifolium* 'October Skies'

∧ *Helenium* 'Mardi Gras'

soften the strong personalities of structural plants, creating visually restful backdrops for the more striking shapes.

Especially interesting are plants that create color themes in natural settings. Think about the yellow theme of American prairies in early fall or the blue theme of a floodplain forest in early spring. *Solidago* and *Mertensia* are strong theme genera and they powerfully bring to mind the larger landscape from which they're derived. The Mixed Planting system developed in Germany has a heavy focus on visual themes like colors to evoke specific habitats. For example, many of the companion species in the mix *Silbersommer* (Silver Summer) are plants from Mediterranean-type habitats, focusing on plants with predominantly silver foliage. Plants in the Flower Steppe mix focus on perennials with subdued violet-blue and yellow flowers, suggestive of the Eurasian steppe. Connecting the colors of your seasonal theme plants with the characteristic colors of a wild community will strengthen your planting's evocative edge.

LAYER 3: GROUND-COVERING PLANTS

The ground cover layer is the essence of a plant community—primarily a functional layer. Once the design has been created with the first two layers, one now fills in between with ground-covering plants. These species may not have the striking forms or pretty flowers of the design layers, but they literally and figuratively hold the community together.

Ground cover plants are typically low woody or herbaceous species that live under or around the base of the design layers. This layer includes plants with aggressive, clonal-spreading behavior, such as *Packera obovata*. They closely hug the ground and provide excellent erosion control, weed suppression, and green mulch function. In grassland communities, this layer can be formed by a thick covering of short grasses and creeping forbs, such as *Glechoma hederacea*, *Packera aurea*, or *Carex pensylvanica*. In woodland communities, the layer can include spring ephemerals, ferns, sedges, and low woody plants of the genera *Vaccinium*, *Calluna*, *Artemisia*, or *Origanum*.

The ground cover layer often receives full sun in spring and early summer. Later in the year, as perennials grow taller, ground cover plants are often partially or fully shaded, which can cause them to go partially summer dormant. They often flower and fruit before this happens and use the available growing window similar to what spring ephemerals do in forest plant communities. Some geophytes from genera like *Galanthus* and *Erianthus* fall in this category. Their large underground storage organs allow them to survive during unfavorable growing conditions.

The purpose of this layer is to achieve the highest possible functionality without compromising the legibility of the design. Providing essential ecological function, such as covering soil or providing pollen sources for insects is equally important as aesthetic quality. Therefore, when designing we are not only looking at year-round aesthetics, but year-round functionality. For example, a rain garden or bioswale needs a full-time erosion control layer to stabilize soil in the dormant season. A pollinator display garden needs continuous nectar sources for insects. As a result, plant selection should respond to the functional need. Consider the following examples.

Storm water management. Select plants with evergreen basal leaves for winter erosion control and evapotranspiration. Choose species with a diversity of root systems, especially deep roots for better rain water percolation. Pack in as many plants as possible to increase water absorption and filtration.

Erosion control. Select evergreen and semi-evergreen species with persistent basal foliage. Choose aggressive clonal and self-seeding species that can vegetate heavily eroded soil on their own.

Soil building. Select legumes such as species of *Dalea*, *Thermopsis*, and *Lupinus*. Work with herbaceous species as much as possible. Their ephemeral root systems store carbon in soil. Use plants with deep roots to bring nutrients up from lower soil horizons.

Phyto-remediation. Use plants with high biomass production and pollutant uptake capabilities, such as species of *Typha*, *Scirpus*, *Panicum*, and *Carex*. Diversity of species is essential to functionality. Try to use a mix of plants with different behaviors, such as clonal spreading or quick seeding. This will encourage them to spread, reseed, and naturalize, closing gaps in the vegetation after disturbance. Focus on semi-evergreen or evergreen species that cover the soil all season long. Change is acceptable in this planting as long as it does not alter the design too much or form a monoculture. Do not be afraid of rhizomatous and stoloniferous species. Highly competitive species can be combined with one another to keep each other in check. These tough plants are essential for plantings with low maintenance or heavy competition from invasives. We need the native equivalents of *Vinca minor*, *Pachysandra terminalis*, and *Hedera helix*. Disturbance-sensitive species may work in a residential setting, but tough sites require resilient plants.

One of the problems with using ground cover plants from native habitats is that many of these species are rarely commercially available. They are just not ornamental enough. Some you can only get by seed. But as designers, we can substitute low, shade-tolerant plants available in nurseries to fill the role of native ground cover species. Commercially available substitutes for the native ground layer include sedges—particularly those that are tolerant of dry shade. Rhizomatous species like *Erigeron pulchellus* and *Meehania cordata* are great for their ability to creep and cover. Long-lived clump perennials like *Heuchera villosa* and *Asarum canadense* stay in place and can live under other species. Plant ground covers wherever there is space for them: under trees, shrubs, and taller perennials. Fill all gaps between taller plants of the design layer. Use them like you would mulch.

⩗ *Carex cherokeensis*

⩗ *Erigeron pulchellus* var. *pulchellus*
   'Lynnhaven Carpet'

∧ *Symphyotrichum ericoides* 'Snow Flurry'

⩗ *Meehania cordata*

⩗ *Carex pensylvanica*

∧ *Asarum canadense*

⩗ *Heuchera longiflora*

⩘ *Callirhoe involucrata*

∧ *Geum fragarioides*

Notice the difference in ground cover and species diversity.

TRADITIONAL PLANTING PLAN

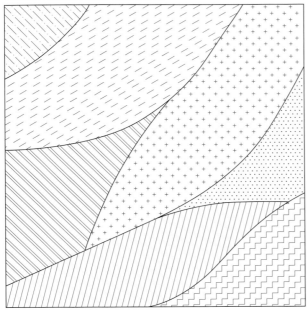

In a traditional planting plan, plants are grouped together in masses of single species.

DESIGNED PLANT COMMUNITY PLAN

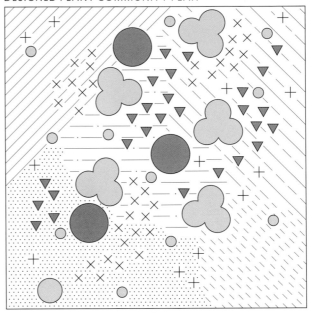

A designed plant community plan creates groups of compatible species that interact with each other and the site.

| structural plants | seasonal theme plants | filler plants | ground cover plants |

LAYER 4: FILLER SPECIES

Because large structural species can take years to establish, use temporary filler species to cover the soil until plants are mature and to provide visual interest. Good filler species reseed on their own and keep a population alive by jumping from gap to gap within planting. They eventually disappear when the design and functional layers are thick enough. Filler plants should comprise approximately 5 to 10 percent of a mix—enough to make the layer visible and build a good seed bank for future spreading. Examples include annuals like *Cosmos* or *Ammi majus*; short-lived perennials like *Achillea, Aquilegia,* and *Knautia macedonia*; and short-lived grasses like *Hordeum jubatum, Nassella tenuissima,* and *Briza maxima*.

Filler species can create seasonal color themes, but since they are dynamic and rather unpredictable, consider them as serendipitous accents rather than mainstays of your design. Examples of good fillers are annual, biennial, and short-lived perennial plants. Species of Grime's ruderal category are ideal because most of them produce huge

⚐ *Agastache rupestris*

⚐ *Aquilegia canadensis*

⌃ *Coreopsis verticillata*

⚐ *Spigelia marilandica*

⚐ *Chrysopsis mariana*

⌃ *Euphorbia corollata*

⚐ *Delphinium exaltatum*

⚐ *Lobelia cardinalis*

⌃ *Silene virginica*

amounts of seeds in their short lifetime. The seed is stored in the seed bank and will produce plants even years after installation if a planting is disturbed and open soil left behind. For this reason, it is good to distribute filler plants evenly throughout your composition, giving them an opportunity to self-seed throughout.

Filler species can function like inbuilt insurance, repairing planting from within after disturbance. One such example is *Lobelia cardinalis* in rain gardens. Once a rain garden's vegetative cover is dense, lobelia usually disappears. It does not tolerate much competition from taller species and is generally short-lived. For many years to come, this plant might not be present in the planting. However, if the plant layer is disturbed by a contractor fixing an under-drain, for example, open soil is left behind and sunlight reaches the dormant seed embedded in the soil. In such cases, lobelia can germinate and magically reappear within a planting until it reaches the end of its short life, or is outcompeted by other perennials.

APPLYING THE LAYERS

The following examples illustrate the process of layering plants in each of the three archetypal landscapes.

Open grassland community

Start by selecting visually dominant, evocative, and aspect-forming species in distilled and amplified patterns for the structural layer. These plants are taller perennials and grasses with clump-forming habit. They are long-lasting and well behaved within a planting. For example, *Baptisia australis*, *Andropogon virginicus*, and *Eutrochium fistulosum*.

Fill gaps between plants of the structural layer with a dense layer of ground-covering forbs and grasses covering the soil. Use preferably evergreen or semi-evergreen species to provide erosion control during dormant season and to suppress weeds. In high-disturbance areas or on sites with very limited management resources, strongly rhizomatous or stoloniferous species are preferred. These can form seasonal themes but are otherwise companion species to those that are taller and more visually dominant. High species and age diversity is possible here because this is not the dominant design layer. Examples include *Carex amphibola*, *Viola sororia*, and *Juncus tenuis*.

Include dynamic temporary species to fill the gaps in the early stages of a planting. These can also form seasonal themes at certain times of the year—for example, *Verbena bonariensis* or cosmos in summer. They may not persist in a planting, or may only come up when open soil emerges, such as through disturbance. Include as many plant morphologies as possible to form a diverse plant community. For example, include bulbs such as species of *Narcissus*, *Crocus*, and *Camassia*.

Make sure if one species fades at certain times of year, another fills the spaces it leaves behind. For example, spring ephemerals can be combined with late-emerging ferns and warm season grasses to prevent gaps.

∨ Structural Layer

∨ Ground Cover Layer

∨ Seasonal Theme Layer

∨ Dynamic Filler Layer

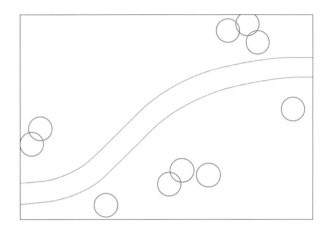

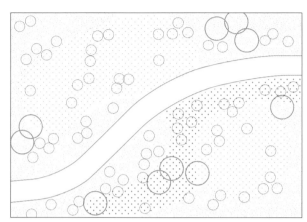

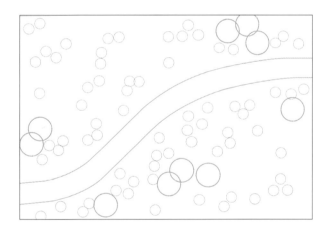

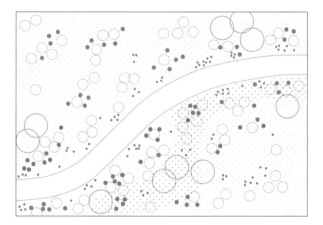

Woodlands and shrublands community

Add visually dominant and seasonal theme-forming trees and shrubs in amplified patterns to the herbaceous plant layers. Fill gaps with a dense layer of ground-covering forbs and grasses. The species of this layer have to tolerate a broad spectrum of light conditions, ranging from full sun to shade. Good choices might be *Carex cherokeensis*, *Deschampsia cespitosa*, and *Erigeron pulchellus* var. *pulchellus* 'Lynnhaven Carpet', which can form seasonal themes at certain times of the year. High diversity is sought in this layer, without sacrificing the clean legibility of the design. For example, several similar species of *Carex*, *Heuchera*, or *Phlox* could intermingle, and no one would notice that they are different from each other. However, the insects would find higher food diversity and the site would most likely have higher ecological value and be more resilient than if it was a monoculture. Intermingling species have to be compatible.

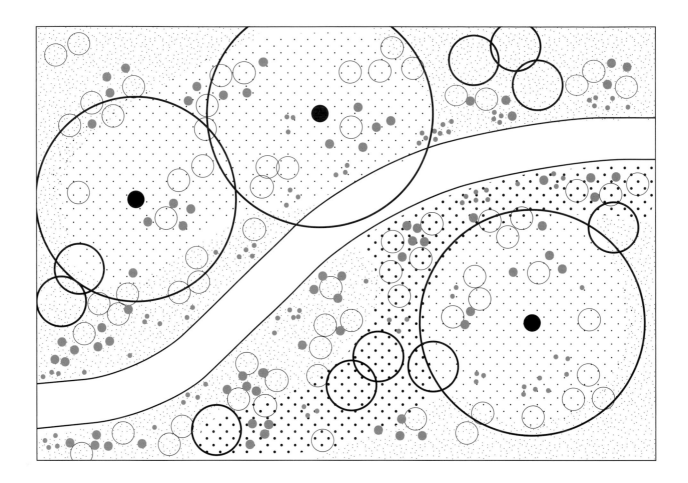

This archetype is already visually interesting without plants of the ground layer. Even the color ranges created by light and shade are spectacular. Often, this type benefits from a slightly more restricted plant palette, so attention can rest on the play of open and closed canopy and the striking patterns created by the mix of woody and herbaceous planting. Too much visual diversity in the plant palette can distract from this unique quality of open woodlands. Keep the complexity in the patterning and spatial composition; each vegetative layer should be very simple.

Open forest community

Establish or enhance a closed tree canopy and visually open space underneath tree trunks. If adding an understory layer, select small trees and tall shrubs that are transparent and open. Start to layer the ground plane with a dense mix of ephemerals, sedges, ferns, and mosses. For the thematic seasonal layer, focus on spring ephemerals and late summer asters and woodland sunflowers. Intermingle with evergreen species, such as *Polystichum acrostichoides* and *Carex plantaginea*.

Many forest landscapes on the East Coast of the United States have been depleted of a rich ground cover layer by an overpopulation of white-tailed deer. The sad truth is that the spring ephemerals everyone associates with forests are frequently missing—the forests feel empty and the lack of ground-covering vegetation is an open invitation for deer-resistant exotics like *Alliaria petiolata* and *Microstegium vimineum*. Deer fencing may be required if you are trying to create an open forest community in this area.

Trees grow slowly and taking excellent care of the one layer that makes this archetype possible is essential. Professional arboricultural care from seedling to full-grown tree is the foundation for a safe and thriving forest plant community. The next generation of trees has to be present in the ground or shrub layer and it has to be protected from deer and damages. Though windfall and ice damage can never be prevented, keeping trees healthy and properly pruned can dramatically reduce the risk of damage. Wind should go through your trees, and double leaders or included bark should be dealt with in the early stages of a tree's life. One should be able to see under trees; limb them up so the area underneath the lowest branches is open.

CREATING AND MANAGING A PLANT COMMUNITY

Any designer who has ever created and installed a planting, walked away, then visited that planting five years later learns that design is not a singular vision set to paper; it is a thousand small decisions and actions continuously made. It is those seemingly small decisions and actions we want to elevate here. Many of the problems that plague traditional horticulture stem from the awkward division of labor that separates designers from growers and installers. We believe that good design emanates from the site, not from a cubicle, and good installation and management follow a big vision, not reactionary decisions in the field.

SITE PREPARATION: THE DESIGN PROCESS CONTINUES

Complex plant communities only persist if designers and land managers collaborate. Here, clipped hedges grow next to diverse plantings, requiring a combination of horticultural and ecological management techniques.

Building a designed plant community differs significantly from traditional planting. It is not that our method introduces a new set of techniques and tools. Rather, our method requires rethinking the why, how, and when of applying these tools. We question the set of widely accepted and blindly applied maintenance techniques precisely because many simply do not work and because they are sometimes applied without a goal in mind. Therefore, we require a rewriting of cookie-cutter installation and maintenance specifications. Of course, all planting relies on altering a site to accommodate plants that would not naturally be there. Our goal is not to discourage intervention, but rather to more thoughtfully align it with natural processes of soil building, plant competition, and ecological succession.

Thinking of your planting as a community that evolves over time pushes the design process beyond your computer screen to the site itself. It requires a rich collaboration with contractors and garden staff, and it encourages long-term engagement with your planting to help it adapt as it matures. The process marries techniques and tools from the ecological and horticultural realms.

TEMPORARY ACTIONS TO ESTABLISH YOUNG PLANTS

Preparing a site for planting extends out of our principle that the constraints of a site—its shade, wet clay, or steep slopes, for example—are actually assets that will aid in creating a unique community of plants. So the goal of site preparation is to preserve these unique qualities, while providing optimal growing conditions.

All practices are driven by ideas, so our starting point is to understand the assumptions that propel flawed procedures. Conventional site preparation seeks to transform a site into a neutral backdrop for a new design. Take, for instance, traditional soil preparation. It focuses on eliminating the distinctive qualities of the soil. Practitioners spread amendments to balance the pH, mix compost to raise the organic matter, and till soil until it is fluffy and soft. The goal is a loose, friable, deeply fertile black soil. While soil like this may be well suited to crops or annuals, it is problematic for many native and naturalistic species. Highly disturbed, fertile soils increase competition, encourage ruderal species like weeds, and perhaps most troubling of all, can decrease the lifespan of many garden plant genera like *Echinacea*, *Salvia*, and *Sedum*, which prefer leaner soils. Landscape soils should not look like potting soil.

During site preparation, extremely poor soil was amended with large amounts of mushroom compost; a waste product of the local mushroom industry. The compost was tilled in and the soil fluffed up. During installation, planters sank in several inches as soil settled under their feet. The high nutrient content and salt levels of the soil led to devastating plant losses within the first growing season.

This problem is compounded by recommendations from soil labs. Don't get us wrong: having soil tested is valuable in learning the conditions with which you're dealing. But most soil test labs recommend balancing nutrients based on a one-size-fits-all standard that assumes all plants like a highly fertile, perfectly balanced mix. The problem with this approach is that plants have evolved to soils with particular textures and chemistry. They do not want some kind of generic, ideal soil; they want specific soil. Members of the heath family love acid soils; the olive family loves alkaline soils. Some plants such as *Ammophila breviligulata* need sandy, low-nutrient soils, while *Rudbeckia laciniata* thrives in nutrient-rich clays. A site can be changed to fit a plant; but only when a plant fits a site will your planting truly be self-sustaining.

The big shift from tradition is in thinking of site preparation as temporary actions needed to establish young plants—versus permanent, one-time-only activities. All sites need minor alternations to help new plants thrive. Consider the massive adjustments plants make moving from a nursery to a site. When nursery plants arrive on site, they are not yet acclimated to the local microclimate and soil conditions, often coming right out of the ideal growing environments of nursery greenhouses. These plants are tender, not hardened off, and in some cases, may not even have been exposed to direct UV light. Their roots are likely surrounded by peat-based soil media with perfect nutrient and pH levels. In order to survive, they will need plenty of space to reduce plant competition, and optimal growing conditions to become established. Especially if plants already exist on site, adjustments must be made to ensure a smooth transition.

Verbena hastata thrives in heavy, wet soils.

Phlox paniculata 'Jeana' prefers consistently mesic and rich soils.

Eupatorium hyssopifolium is at home in sandy coastal soils.

> Most plants are grown inside greenhouses, under optimal conditions, to encourage healthy plant development and prevent pests and diseases.

SITE STABILITY IS ESSENTIAL TO LONG-LASTING PLANTINGS

Only stable site conditions sustain stable planting. Think about the set elements of a landscape—its geology, soils, climate, and existing structural plants. These elements naturally support a group of corresponding plants; it is our responsibility to select the right ones. It is all too tempting to rely on a narrow group of plants we know, but the drawback of forcing a plant list on a site is that it requires costly alterations to the soils, most of which do not last. If a site's soil and underlying bedrock is too alkaline for a bog or heath plant community, for example, amending with peat moss or finely ground sulfur may improve the soil for a few years—however, peat will decompose and underlying limestone will eventually raise the pH again. The same can happen if a planting site is located between urban concrete structures. Amending with acidic substances will bring the pH down for a time, but runoff from concrete can raise it again. Without continued amendments, the desired plant community will likely not be able to survive over the long haul.

Also keep in mind that too many landscape architectural specifications rely on engineered soil recipes that list components like an ingredient list, particularly for storm water management facilities. Like the one-size-for-all recipes for planting, none of these engineered mixes function like natural soil. They tend to focus exclusively on some functional characteristic of the soil, like its bearing capacity under sidewalks or how quickly water moves through it. These factors are important, but what is lost in these equations is an understanding of how the mineral substrate interacts with the living organisms of roots and microbial life.

Understanding your existing soil will result in informed plant selection. Have your soil professionally tested. Collect multiple soil samples from different areas of your site and send them to a trusted soil lab. Take care to correctly interpret soil test results. Most labs will evaluate soil test results for you, but as mentioned previously, be wary of

<< Digging a soil test pit is the easiest way to understand a site's soil conditions. Narrow spates are excellent for this job.

< Feel the soil with your fingers to learn about its water-holding capacity, organic matter content, and level of compaction. This soil has fantastic structure and its dark color is an indicator of high organic matter content.

their recommended amendments. Soil labs are not plant experts and in some cases, recommended amendments do more harm than good. Even better than relying on soil labs is to consult with a university soil scientist to read and interpret your soil tests for your specific site and goal community.

If the soil does require amendments for the indicated plant community, make sure to apply them at the correct time of year. Carefully follow dosage instructions—more is not always better and can make things worse: more nitrogen than recommended can cause plants to grow taller and become heavier than their frames can support, resulting in floppy, unsightly plants. Worse still, if a soil exceeds its capacity to uptake nutrients, excess quantities will run off with rain, polluting rivers and streams. Minimize soil amendments within storm water management plantings in particular, because such runoff contains high levels of nitrogen and phosphorus, so plants do not usually need more. Plants take more nutrients from runoff if a soil is lean.

Sometimes plants just need a little help at the beginning, and we can kick start their development by adding organic matter or specific nutrients, if needed. If a soil is low on organic matter or completely constructed, initial soil conditioners can be beneficial for plant development. Compost tea and light top dressing with compost are sensitive amendment techniques that help plants get established without dramatically altering a site's conditions.

Use amendments in careful combination with appropriate plants to avoid the need for soil adjustments beyond establishment. Plant debris is constantly broken down by microorganisms and essential nutrients are released back into the soil. Encouraging this activity in your soil is simply a matter of allowing the natural process to work. For designers, the big takeaway is that no product will improve poor plant selection. Just like no vitamin tablet will ever replace eating your vegetables, nothing replaces good plant selection and the natural cycle of healthy soils.

Plants are an essential player in building soil. A perennial's deciduous root system, its underground storage organs, and a legume's ability to store nitrogen, just to name a few examples, are the best and most sustainable soil amendments one can find. Every fall and winter a large percentage of a perennial's root system dies off. The roots leave behind empty channels as well as organic matter in a stable form called humus. This is the way carbon and nutrients are stored in a soil. Hundreds of thousands of root channels will heal and rebuild even highly disturbed and compacted soils over time, and enrich low-lying soil horizons with organic matter. The more roots, the more quickly a soil is restored. In order to get as many roots in the ground as possible, plant as densely as possible and use a diversity of root morphologies to interact with the soil at different levels. Successful designed plant communities do exactly that—combine different shapes with one another in order to achieve the highest possible density. Every inch above- and belowground is filled with plants.

Dense growth provided by these species of *Iris*, *Equisetum*, *Onoclea*, and other perennials actively builds soil in James Golden's garden.

Engineered soils are typical for storm water management plantings. More than 60 percent sand, these media replicate coastal site conditions. That said, many traditional storm water management species do not tolerate such soil, while coastal species can often be successfully established.

< This rain garden was amended by gently raking in a few bags of organic soil conditioner, instead of massive amounts of nutrient-rich compost. The goal is to create better growing conditions for young plants, without altering site conditions too much. The conditioner will mix with the soil during plant installation.

Temporary fencing and signage protect this recently planted bioswale from mowing crews cutting the adjacent turf areas, training maintenance workers in which areas need mowing and which do not.

LIMIT THE EXTENT AND AREA OF DISTURBANCE

Disturbance creates the need for management. One of the greatest threats a young planting faces is invasion from weed species. Many weeds are hemerophiles, that is, plants that thrive in habitats disturbed by humans. Site preparation and planting disturbs a site, creating ideal conditions for these species to germinate. Depending on the time of year, a young planting may be invaded with weeds like spotted sandmat, crabgrass, and horsetail. Minimizing the area of disturbance is the best strategy for preventing weed outbreaks. The less you disturb, the less you have to revegetate, meaning less management is required and fewer resources are needed.

It's easy to focus on the planting area itself, but don't neglect to properly stabilize and protect other parts of the site as well. Every act of disturbance, however small, attracts invaders that could damage new plantings. Even stockpiling construction materials for a few weeks might create a bare spot that becomes an ideal habitat for garlic mustard or Japanese stilt grass. In order to prevent weed outbreaks like this from seeding into planted areas, cover disturbed soil with vegetation as soon as construction is finished. Seeding bare spots with quick-establishing grasses or forbs helps to cover the ground and prevent invasions from occurring.

Protect soils and existing vegetation by fencing them off during (and, if necessary, after) construction. The best way to minimize disturbance and soil compaction is to never enter certain zones of a site. Cover future planting areas with sheets of plywood, temporary applications of mulch, or thick geo-textile mats in order to keep equipment from compacting a soil during construction. This method distributes the weight of a machine over a larger footprint and prevents soil disturbance. It may seem difficult to stay within construction limits, but the consequences of disturbance are difficult to repair. Damaged trees, compacted soils, and invasive species can spoil almost all the benefits of a planting.

CLEARING A SITE

Taking the time and effort to properly remove undesirable vegetation will save you immensely in time spent managing a site later. Newly installed plants are vulnerable. They lack the root system and resources to compete with other aggressive plants. Existing weeds can deprive intended species of nutrients and water, and they can crowd out young transplants because of their height and larger leaf mass. Therefore, limit competition from existing plants during the establishment phase. Start by identifying any potentially invasive species on your site. Nearby stands of notorious invaders like small carpet grass, Oriental bittersweet, or Canada thistle should be red flags when you visit a site for the first time. If you see adult specimens, there is almost certainly plenty of their seed stored in the underlying seed bank for many years to come. It might make sense to replace the top layer of soil with clean topsoil, or to add clean soil on top of the existing grade to prevent a maintenance nightmare.

Learn Your Weeds

—

In order to write correct weed removal and long-term control specifications, designers have to be familiar with the problem species or bring a specialist on board. Purchase a weed identification book or smart phone app, and have it with you whenever you go out on a job site.

Avoid heavy equipment whenever possible; big machinery often fakes efficiency. Rain gardens are commonly dug out by backhoes and fine grading is done by skid loaders. Inappropriate or oversized equipment does more damage than good, and it takes years for a site to recover from the compaction and unnecessary disturbance. In the time spent waiting for machinery to be delivered to a site, a team of five could have prepared a planting area with three rakes, two shovels, and no compaction.

In addition to functional problems, weedy sites also create an image problem for young plantings. We often associate weeds with feral fields and unkempt grounds. They convey an image of neglect. This is particularly troublesome because young mixed plantings often lack the structure and flowers of mature species, making them hard for those not horticulturally inclined to distinguish from weeds. When the public cannot distinguish a new planting site from a weedy field, it can create a vicious cycle of disinvestment. Attractive, well-maintained plantings do just the opposite, attracting people and encouraging even more care.

Remove all parts of a weed, including its roots and other underground storage organs. Take particular care to either remove or keep dormant the bank of seeds stored in the soil, if possible. Do not underestimate the volume of weed seeds that may be banked in your soil. Thousands of seeds can exist in a mere square foot, accumulating for decades and lying dormant until the opportune moment. Target weeds at times when they are most vulnerable. Every weed species has a weak spot in its life cycle—this is when we need to take action. Often the best time to spray or mow is when weeds are just emerging or newly mature, prior to setting seed. This requires you to understand a bit about the lifecycle of weeds. Some, like Japanese stilt grass, are annuals. It would be a waste of time and resources to spray them with an herbicide in early October just after

⋏ Heavy equipment can cause compaction and deep tire ruts.

⋏ Even if a soil is tilled after grading, it can take plants and their root systems many decades to restore pre-construction conditions.

⋏ Hairy crabgrass (*Digitaria sanguinea*) emerges in mid-May, threatening to smother young transplants. The weed outbreak could have been prevented by adding a few inches of clean topsoil before installation.

⋏ Where possible, burning unwanted brush is a highly effective site preparation tool, leaving behind a generally clean slate for planting, limiting soil disturbance, and recycling plant essential nutrients.

∨ Undesired species such as *Trifolium pratense*, *Lotus corniculatus*, and *Setaria faberi* have overrun this site. Small plants have a hard time competing and the area looks more like a fallow field than a designed planting.

∨ The right tools matter. Deeply rooted undesirables must be removed completely—leaving a small piece of root in the ground will allow them to grow back.

their seeds have ripened. They will die back with the first frost. Fast-growing rhizomatous perennials like Japanese knotweeds are best treated in early spring, while their leaf mass is small and spraying can be concentrated at the base of the plants. If your site has a particular issue with undesirable vegetation, you may have to devote an entire growing season (possibly even two) to ridding your site of weeds.

Choose a weed removal strategy best suited to your site and resources. Various approaches exist, ranging from fast techniques that require chemicals or cultivation, to soft techniques that take longer. Tailor your strategy to a site, and consider several techniques in combination with one another to deal with different kinds of weeds. In new plantings, broad techniques work. In enhancement plantings, more targeted techniques are necessary in order to protect desired species and cause as little disturbance as possible. The following chart gives an overview of most available weed removal techniques.

WEED REMOVAL TECHNIQUES

WEED REMOVAL TOOLS	MATERIALS	BENEFITS AND CHALLENGES
Smothering	Recycled paper and cardboard Organic mulch (bark, wood chips, compost) Clean topsoil	Difficult in enhancement planting Ideal for container planting, not seeding Safe to use around existing trees and shrubs if thin layer is applied Little impact on soil health because rain and air pass through materials Requires long lead times in order to kill off deeply rooted species
Spraying	Organic herbicides Traditional herbicides	For enhancement planting (spot-spraying) or new planting Some herbicides can be harmful to people and the environment
Mechanical removal	Hand weeding Machinery (brush hog, string trimmer)	For enhancement planting Manual or with machines Causes high levels of disturbance
Burning	Propane burner Drip torch	For enhancement planting Burned debris makes plant essential nutrients available immediately for other plants Selectively reduces pressure from fire-intolerant species like cool season grasses and winter weeds
Cover cropping (competitive exclusion)	Seed	Requires long lead times Cover crops can be part of the future design Can enrich and improve soil (legumes enrich soil with nitrogen) Temporary solutions to bridge time between site preparation and planting

This soil is severely compacted by human foot traffic—a common problem in populated landscapes.

Many years of parking use led to highly compacted soil here. A pickax and spade were needed to dig the test pit, which revealed a shallow soil horizon.

Compacted sites under mature trees should not be tilled, to avoid damaging sensitive surface roots. Here, augers were used to plant landscape plugs.

TREAT COMPACTED SOILS BEFORE PLANTING

Almost all sites we encounter in urban and suburban environments suffer from some form of soil compaction caused by construction, site preparation, runoff, or human use. In fact, compaction is so prevalent and damaging to planting that it is astonishing how few design specifications address how to deal with it.

Compaction prevents rain and irrigation water from penetrating a soil, causing runoff and preventing a plant's roots from getting moisture—even if water is plentiful. The anaerobic conditions caused by compaction make difficult habitats for microbes. Some plant species depend on these microbes and live in symbiosis with them. As a result, if soil microbiology is unhealthy, not all plant species will be successful. If a soil's bulk density is above a certain limit, plants cannot push their roots deep enough to reach moisture they need during hot weather. When soil is compacted to this level, plants need our help.

There are several ways of identifying compaction. Some are highly accurate yet cumbersome and others are simple yet not terribly rigorous. Pick a method that fits best with your site and budget. Start with the easiest first. If your project requires more verifiable techniques, use those next.

Tilling may be a designer's first instinct when dealing with compaction, but it can compound soil problems. While tilling a soil may break up compaction in the upper few inches of soil, it does not penetrate deeper. Compaction often happens several inches to even a foot or two below the surface. Tilling loosens and mixes soil, but a loose soil is not automatically a good soil. Gardeners have long had a romance with the idea of

METHOD FOR IDENTIFYING COMPACTION	DESCRIPTION	LEVEL THAT RESTRICTS ROOT GROWTH	ADVANTAGES/DISADVANTAGES OF METHOD
Metal Stake Test	Any kind of straight stake that can be pushed into the ground.	If you cannot push stake six to eighteen inches into the ground.	An easy technique for identifying potentially compacted areas. Stake should move through uncompacted soils relatively easily. Not scientific or measurable.
Cone Penetrometer	A simple metal stake with a gauge on it. As you press stake into the soil, it measures resistance in psi.	160 to 300 psi	A coarse gauge of compaction level that simulates the movement of roots. Relatively affordable; easy to use and can take multiple samples. Penetrometers do not measure pores in soil (freeze/thaw, earthworms) which roots can move through.
Bulk Density Test	Bulk density is a measurement of the weight of a soil in a given volume. Bulk density increases with compaction.	approximately 1.6g/cm^3	One of the most reliable measures of compaction. Difficult to test; requires special equipment and an oven. Complicated testing method; taking lots of samples is difficult and leaves room for human error.

loose, well-tilled soil as the ultimate planting medium. The problem is that tilling collapses many of the pore spaces in soil, ultimately causing soils to settle. Planting directly into fluffed-up soil causes roots and sensitive plant crowns to become exposed after a few weeks, as water settles and reduces the soil's air gaps. As mentioned, tilling can also destroy soil's microbial networks and bring up dormant weed seed from lower soil horizons. A great alternative to tilling is deep plowing or sub-soiling, which loosens up compacted soil horizons without destroying soil structure. Unlike a tiller, deep plowing preserves the natural horizons of soil, while creating vertical channels for air, water, and roots to move through. Deep plowing depth can be set to where the compacted soil layer is located. For small urban sites where a plow is not feasible, consider using a walk-behind vertical trencher or core aerator.

Soils will not naturally recover from severe compaction; if water and roots cannot move through a hardpan layer, the compaction can persist for centuries. However, plants and the natural processes involved in soil evolution—freezing/thawing, moisture/drying, earthworm burrowing—can mitigate moderately compacted soils. Over time, roots can break through and loosen up crusts and lenses. A designed plant community accelerates this process because it provides a diversity of root types to help penetrate soils at different depths. Each new root is a small drill that opens up gaps. The diversity of layered plants provides an important ecological function not offered by traditional planting.

Preparing a site for planting is all about setting the stage so that natural processes of root growth and soil building take place. It is about understanding soil as a living partner with plants, not some inert material that we must break into submission. It is also about managing competition early, sheltering your plants from the mob of aggressive species waiting to colonize disturbed ground. Neglecting this process will compound site problems later on; doing it right up front may just reward you with years of lush planting.

Proper site preparation sets the stage for lush vegetation like this designed forest plant community.

INSTALLATION: USING A PLANT'S NATURAL GROWTH CYCLE TO YOUR ADVANTAGE

A community-based approach to plant installation differs from conventional planting in a couple of important ways. One, installation practices are based on plants' natural establishment rhythms, not project deadlines and opening events. Because we focus on a mix of plants, each with different metabolisms and competitive strategies, our approach takes the timing of installation seriously, aligning the project schedule with a plant's optimum rooting times. Second, our approach extends from the principle that plantings are vertically layered. As a result, we install plants in layers, making sure the design gestures and functional relationships of each layer are clear at the time of planting.

This elevates the role of the designer during installation. Too often during construction, project needs come before plant needs. A brilliantly conceived planting may bleed a slow death from a thousand little nicks of schedule changes, budget cuts, and poor plant storage, selection, and installation. Designers understand

Installing a designed plant community for a storm water management system in Lancaster, Pennsylvania.

Less than a year after installation, nearly 100 percent of all small plants have survived the transition from nursery to this harsh urban landscape in Baltimore, Maryland.

the frustration of watching a construction schedule slip from April into the dead heat of summer. Or seeing an agonized-over plant palette dissolve as a contractor substitutes half the list for more conveniently available plants. Designed plant communities require special advocates during installation; this requires not only better specifications, but also on-site guidance and layout.

Great installation practices result in 100 percent transplant success. The goal is for every transplant to live and for all seeds to germinate at first attempt. If this is achieved, a planting will be truly sustainable, limiting the reliance on replacement plants, fertilizer, and continuous irrigation.

TIMING FOR OPTIMUM PLANT ESTABLISHMENT

Installation must follow soon after a site has been prepared for planting. Much of this strategy is to prevent weeds from invading open soil. But the other threat is soil degradation. When soil is exposed to sunlight, rain, and extreme temperature changes, it damages the sensitive balance of microbiology and nutrients. The carbon stored in soils is the main component of organic matter, and thus an essential source of nutrients for plants. It is what makes soil functional from a vegetative point of view, supporting vital aspects such as water-retention capacity, structure, and fertility. When soil is exposed to sunlight and air, the carbon oxidizes in the form of carbon dioxide. Frequently cultivated soils, including farmlands all over the globe, have lost 50 to 70 percent of their original stock of carbon—one of the reasons climate researchers focus so heavily on regenerative agricultural practices. The longer a soil is exposed, the harder it is to vegetate later. To get a site growing again quickly, move ahead fast, using as many plants as possible to establish dense vegetative cover.

If project schedules do not line up with optimal planting windows, consider a cover crop to bridge the time between site preparation and plant installation. Cover crops are cheap and easy to establish. In most cases, they are grown from seed which can easily be broadcast over a site right after preparation is finished. Different types of clover, peas, and vetch are leguminous, fixing nitrogen back to the soil. Nonlegumes such as annual rye grass can function as nurse crops, covering the soil and taking up excess nitrogen in the soil. Annual nurse crops can be warm or cool season growers, providing designers a wide range of options for different times of year. Choose cover or nurse crops based on your goals. If the objective is to restore fertility and microbial activity to a soil, legumes may be a good choice. If the goal is preventing erosion, annual grasses may work better. Consider your timeframe as well. Plants such as buckwheat, oats, and radish germinate quickly and die back readily in winter; other plants like clovers are slower to establish, making them poor short-term crops. Consult with a meadow specialist or seed provider to select the best species for your project.

Cover crops will not work for all projects or timeframes. Small, urban projects with only a few weeks of open soil should consider temporary mulches to keep soil covered. Remove excess mulch prior to planting. Midwinter preparation is another instance in which a cover crop is not appropriate; in this case, a light mulching with organic material (such as shredded leaves) protects your soil until the time is right for plant installation.

The single most important timing consideration is to make every effort to install plants when they are actively growing. This results in the highest plant survival rates, though it also narrows the optimal planting window from the time after the last frost of spring to a few weeks before the first frost of fall. If transplants are installed during favorable times, watering can be reduced to a bare minimum and plant losses are much lower. The further an installation happens from a plant's optimal planting window, the more watering and help it will need to get established. Ideally, plants should be installed during times of year when rains are more frequent to minimize the need for irrigation.

One week after seed was broadcast, a cover crop mix of legumes and deeply rooted radishes is beginning to establish on this site, bridging the time between final site preparation in early summer and installation of a designed plant community in fall.

Designed plant communities include species of various metabolisms and life cycles, such as warm and cool season grasses, annuals and biennials, and spring ephemerals. This range of life cycles differs dramatically from a conventional block planting of a single species. In a designed plant community, some species in your design might be in active growth during installation, while others are not. For example, warm season grasses are in their prime during late summer installation, but the planting's spring ephemerals may be completely dormant at this time of year. Species of *Narcissus*, *Erythronium*, and *Mertensia* might not have any foliage at all. Most woodland perennials are fully leafed-out by May, but some meadow plants or annuals wait until summer temperatures heat up to actively grow. For most projects, it is not feasible to plant each different plant type weeks or even months apart. Instead, designers should plant various forms of propagules, such as containers, bare root, bulbs, live stakes, seed, and cuttings. Each product has its own installation requirements for successful plant establishment. You may be required to sequence different types in phases, perennials in the late spring and bulbs later in the fall. Reach out to your local plant professional, nursery grower, or seed supplier early in the process if you are not familiar with the correct protocol, to make sure all transplants survive.

While each plant's preferred establishment time may vary, designers should focus on the optimum time of year for most plants. The chart below illustrates the various growth cycles for different kinds of plants. Despite the diversity of timeframes, midspring to early summer and early to midautumn tend to support the widest range of plants. Target the bulk of the planting for these times. Special categories of propagules like bare root plants or bulbs can be added later, during their ideal establishment timeframes.

Optimal installation times depend on a plant's metabolism, morphology, and planting method. This chart gives a rough overview of how installation times vary by plant.

PLANT TYPE	JAN	FEB	MAR	APR	MAY	JUN	JULY	AUG	SEP	OCT	NOV	DEC
warm season plants					■			■	■			
cool season plants				■	■	■			■	■		
spring ephemerals			■					bare root	bare root	bare root		
trees and shrubs	■	■	■							■	■	■
seed mixes	■	■	■						■	■	■	■

dormant, bare, or evergreen foliage present (JAN–MAR) · leafing out, spring foliage emerging (APR–MAY) · full summer foliage, often in flower and setting seed (JUN–AUG) · emerging winter foliage (SEP–OCT) · going winter dormant, foliage dying off, evergreen leaves partially dropping (NOV–DEC)

■ indicates optimal installation window

BIGGER IS NOT ALWAYS BEST: CHOOSING PLANTS

Traditional landscape architectural specifications often focus on selecting the biggest, fullest-looking specimens possible. Plant lists call out large-caliper trees while specifications require full, rounded forms with heavy top growth. The bigger the project budget, the larger the size of plants requested. In fact, the entire practice of specification is fixated on aesthetic qualifications, seeking perfect individual specimens. These standards are based in a culture that rewards instant change and transformation—but real landscapes are slow grown.

Of course, a focus on high-quality materials is certainly a good thing. The problem is that traditional specifications too often equate quality with size and fullness, and too many designers and contractors mistakenly believe that plants installed from large containers will create a more established landscape. Establishment is not a product of a plant's size, but of its successful rooting in the native soil. The idea of placing a fully mature tree or perennial flower into a foreign soil deprives the plant of the chance to develop in that soil. Transplanting also damages taproots; many containers are too

This planting by Adam Woodruff combines species with different metabolisms and underground morphologies. During planting, various container sizes and bulbs were required to account for the different developmental stages.

shallow for their proper development. Deeply rooted species of genera such as *Carya*, *Platycodon*, and *Baptisia* are highly susceptible to this.

The truth is, plants that grow and interact with a given soil from an early age are often the most long lasting, healthy, and resilient. A two-inch-caliper oak tree, for example, will likely surpass a six-inch-caliper specimen within five to seven years. This is due in part to the severe post-transplant stress large plants suffer. One study demonstrated that in a typical tree transplant, only 55 percent of the total surface area of roots was retained during transplant, a drastic loss of the plant's basic infrastructure.

In containers, roots get spoiled by the pampering peat-based soils in which they are grown (sometimes called "candy" soil), and many transplants placed in less-fertile soils are unhappy to leave the container mix. This is particularly a problem if a site's soil is poor or has a low amount of organic matter. As a designer, have you ever been able to pop a dead plant right out of the soil, even years after it was supposedly established? The plant's roots never reached out into the surrounding soil, and died when they did not get enough supplemental water to perpetuate container conditions. Using smaller plants and washing or shaking the peat mix off the root systems before planting will help avoid this problem, forcing plants to push their roots into the surrounding soil to survive.

There are other practical benefits to smaller plants. They do less damage to the existing fibrous root system of trees, when a landscape is installed under an established canopy. Smaller plant size also saves time and money, especially with dense planting, reducing plant costs, installation labor, and shipping and handling costs. The use of packing materials and controversial peat-based soil media is minimized as well. For all these reasons, there has been some shift to using smaller plant sizes in recent years.

Handheld augers are fast and efficient tools for planting landscape plugs.

A good alternative to gallon-sized containers is the use of landscape plugs. These are plants with long, deep roots, grown in trays of typically thirty or more. Conventional liner plugs are shallower and designed to be grown in larger-sized containers before planting in the ground. The deep roots of landscape plugs, however, are designed to be planted directly in the ground. They can be quickly installed with a handheld auger. One person can plant over fifty landscape plugs per hour, compared to only a handful of one-gallon-sized containers. In contrast to the months of watering typically required for large-sized transplants, landscape plugs often need irrigation for only a few weeks.

There are times when larger container plants are appropriate. Many of the longer-lived structural species in an herbaceous layer are particularly slow growing.

Plants like *Amsonia hubrichtii*, *Baptisia australis*, and *Asclepias tuberosa* benefit from being grown in larger-sized containers. *Amsonia hubrichtii*, for example, can take up to three years to reach maturity. Many clients are not willing to wait that long, so using a more mature, gallon-sized container can eliminate a few years of establishment.

In almost every community, there is a mix of plants that are quick-establishing yet short-lived, and slow-establishing but long-lived. Both groups of plants play valuable roles. Fast-establishing plants cover the soil quickly and create the conditions for stability, while longer-lived plants eventually become the backbone of a community, helping it to endure. When we plant all these various plants at once, we have to make sure the more aggressive, quick-establishing plants do not outcompete or smother the slow-establishing species. Using plants of different ages—and different container sizes—is one strategy to balance the varied establishment timeframes. A planting may conceivably be composed of gallon-sized plants for widely spaced, slow-growing structural plants; quart-sized and plugs for the vast majority of theme and ground cover layers; and seed for fast-germinating filler plants. Slightly larger sized material for plants in the design layer also helps make a planting more legible while the rest of the layers fill in.

Landscape plugs with five-inch-deep root systems and guides that prevent root circling are ideal. Note the balance between above- and belowground biomass.

When selecting plants, focus on those that will transition well into your landscape. Do not get distracted by flowers or lush foliage. Many wholesale nurseries grow plants specifically for retail markets, not just landscape installations. The goal of propagation is often to produce heavy foliage and flowers for retail sales, not always to develop deep root systems. Perfect growing conditions in modern greenhouses allow plants to thrive with underdeveloped roots. Only healthy roots and hardened-off foliage in proper ratio with healthy root systems can support a plant in the landscape and make installation successful. Check for containers using root guide technologies that prevent root circling. Make sure plants are hardened off before they arrive on site. Some species, such as *Lobelia cardinalis*, need one vernalization before they flower. Ask your nursery to custom grow crops for specific installation times of a project. This will allow the nursery to harden plants off before they get planted in the landscape.

Choose biodiverse and resilient crops; plants grown from diverse seed sources are usually more resilient and will form a correspondingly diverse population in your landscape. Ask your nursery professional for the propagation method applied to certain species. Seed-propagated species usually have the highest biodiversity if seed is collected from large and varied populations. Many plants are grown vegetatively from tissue culture or cuttings. While this guarantees the ornamental characteristics of the parent plant, it reduces the overall genetic diversity of a crop.

Visit nurseries to see propagation techniques for yourself. If that is not possible, ask nurseries to send you images of the plants reserved for your project. Most nursery professionals are eager to share images of their crops, including plant root systems. Specify exact root dimensions on your plan. Container sizes vary from nursery to nursery, leading to confusion and inaccuracies during the bidding process. Some supposedly one-gallon containers may only contain three-quarters of a gallon of soil media for the same cost. Use standardized container classification systems, such as the American National Standards Institute's SP numbers for correct specification of container sizes.

LAYING OUT DESIGNED PLANT COMMUNITIES

The disconnect between designers and what ends up happening on the job site is one of the main reasons for project failure. The best person to arrange the plants on site is you, the designer. You developed a design and have a vision for the site. You must be there to make sure inappropriate substitutions are not made, to verify quantities, and to layout the plants. Make sure you build the necessary time into your cost schedule, so you can be on site for plant layout and installation.

Plant spacing

Despite many beliefs to the contrary, the size of a container does not change the optimal plant spacing and quantity. Whether a plant starts off as a tiny plug or in a three-gallon container, it will eventually grow to the same width. Having large-sized initial plants does not mean they can be spaced far apart. Under-vegetating a planting can have disastrous results. Do not try to save money by cutting plants and spacing them farther apart. Instead, use smaller plants and seed, or reduce the overall planting area.

Use mature plant size as the basis for plant spacing and quantity calculations. Remember, the density of plants that is the hallmark of a plant community is achieved not by cramming plants tightly together, but by creating several vertical layers within a planting. Each layer has its own spacing, based on a plant's sociability, behavior, and mature size. Be particularly careful of cookie-cutter plant-spacing recipes. Many Internet sources suggest spacing based on traditional horticultural plant combinations—leaving plants way too far apart. We are creating something very different, and our plant spacing has to be understood in layers. The overall plant spacing may be eight to twelve inches on center. However, this spacing reflects the combination of matching above- and belowground morphologies in order to keep plants from outcompeting each other. For example, spacing *Panicum virgatum* on ten-inch centers is too dense and will restrict optimal growth. Instead, space *P. virgatum* on thirty-six-inch centers, with a low ground cover in between. The average plant spacing of both layers is closer to ten inches on center, but each of the separate layers are spaced according to their mature size.

PLANT SPACING

Spacing is based on the mature width of plants as well as their vigor and growing behavior. In general, tall species are spaced farther apart than ground covers.

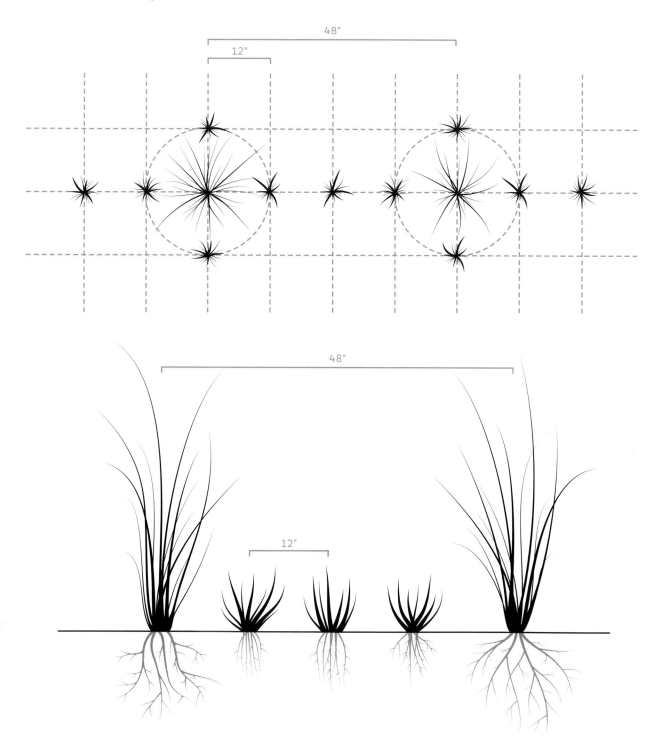

214

Structural and seasonal theme plants can be planted after layout to prevent roots from drying out. Moving and adjusting them will not be possible once they are installed. If more time is needed to make final adjustments, plants can be watered during the layout process to keep root systems moist.

Author Claudia West lays out ground cover species between previously arranged plants.

There are a few special circumstances where tighter spacing is merited. Highly erodible sites or sites that are under extreme weed pressure might require denser plant spacing in order to stabilize the site quickly. Or perhaps you have a very impatient client who is looking for a lush landscape as soon as possible. If this is the case, selective thinning may be necessary. Extremely low-maintenance budgets may call for more plants up front as well. Projects that lack maintenance for weeding or watering may need to more quickly cover the soil.

Laying out plants in layers

Because a designed plant community is conceived in layers, it should also be laid out in layers. Planting plans should be separated or color coded by layers; if not, prepare a plant installation guide prior to planting. It is easier if there are two or three sets of drawings, each representing one of the design layers. The first plan shows the exact location of structural and frame species, the second plan shows the sweeps and drifts of seasonal theme plants. The last layer specifies which mixes of ground covers are to be planted under and between all other layers.

Step 1: Carefully place structural plants. Lay out the plants of this layer mindfully, to make sure the design layer is working well. Such plants are usually the species in the design with lower plant quantities.

Step 2: Arrange seasonal theme plants. You can be less careful in this step than with the first layer. Since seasonal theme plants are usually planted in higher quantities, individual plant spacing becomes less important. Instead, think of bold sweeps and drifts.

After all plants are arranged on site, go back and make adjustments to the layout if necessary. This step is the last opportunity to adjust composition and fix mistakes before planting.

Dynamic plants can also be seeded into gaps after installation. *Rudbeckia hirta* overwintered as seed and germinates between plugged *Rudbeckia laciniata* by early May.

Step 3: Fill in with ground-covering, base-layer plants. Think of this more like putting a population in place instead of placing individual plants. It is not important if these plants are perfectly arranged and planted in a rectangular grid because they are supposed to form a solid ground cover and should reseed or spread clonally. All the extra labor of carefully arranging them will most likely not be visible after two years. Instead of arranging them in a grid, focus on laying them out evenly and densely enough. After they are in place, walk through the site again to make sure there are no gaps in this layer that the plants can't fill on their own within one to two growing seasons.

Step 4: Fill in dynamic and temporary species; add bulbs. Place these species strategically and in between all other layers. Temporary species are meant to disappear after one to a few years; they simply help cover the ground until longer-lived perennials have reached mature width.

Bulbs may be used to create a seasonal theme early in spring, and have to work with the type of ground cover selected for a site. Keep in mind that they often do not directly compete with the other layers because of their metabolism and morphology. Therefore, they can be arranged somewhat independently from other layers.

A planting plan is only a guide

Real design happens in the field. Take time there to get the layout right. Arrange all plants first, then go back and adjust location and spacing. Do not let installation crews, volunteers, or anxious homeowners rush you. Let the contractor know that all plants must be laid out and adjusted prior to the crews coming in—they may want to have a smaller crew on site while you do the layout. Site layout of designed plant communities is much more complex than traditional monocultural massing. The end result often looks confusing to clients and installers. All of the small plants look similar, so the patterns and design gestures are rarely visible at this point. Take the time after layout to explain the design intent to your client. Help them understand what will happen during establishment and how the different layers will become visible over time. A post-installation conversation with your client is highly valuable and reassuring.

Once plants are laid out, work with installers to get the transplants into the ground as quickly as possible, to shade and cool the root systems. Transplants dry out quickly if they sit exposed on a site. In the haste, however, make sure to install plants correctly and with care.

EFFICIENT AND SUCCESSFUL PLANTING

Many container trees and shrubs are severely root-bound when they arrive on a site. De-tangle these roots carefully either by making vertical scores on the side of the root ball, or better yet, by using a metal hook to loosen and pull apart roots. Perennials have more or less deciduous root systems and a large portion of the roots die off every year. If you are planting root-bound containers in the fall, little de-tangling is necessary.

Air is a natural root barrier. If air pockets remain in the soil after planting, plants will take longer to get established and may dry out more quickly. Train the installers to take loose soil around root balls and press it firmly into gaps during planting. This technique will eliminate larger air gaps. Smaller air gaps are filled by thorough hand watering after a planting is installed. The first irrigation is less about keeping plants moist than it is about filling these gaps with sediment. Overhead sprinklers hardly ever do this correctly—they simply soak the soil. Hand watering will take longer, but with the correct nozzle and water pressure, sediment can be washed in to fill the air gaps. If this is done correctly, plants will root and establish faster, dramatically saving time and money.

The edges of planted areas need extra care and special attention, as they are part of the orderly frame of a planting, and highly visible. A planting's edges very clearly communicate the level of care it receives. Therefore, it is essential to work cleanly along these borders. Lay out edges first to ensure they are done correctly before the rest of the area is filled with plants. As with the rest of the planting, mature plant width must be taken into account. Especially if a planting is surrounded by turf, perimeters should be planted more densely or with highly competitive species, to keep invading turf grasses out of the planting. Use neatly overhanging species to cover curbs and structural edges. Place taller species far enough from borders so they never lean into pathways or outside the planting bed, should they flop over after a strong rain or windstorm. This simple rule will prevent aesthetic issues and save on maintenance later.

Additional planting tips that make a difference

Planting in tilled soils. Tilling can fluff up a soil too much for planting. If plants are installed into highly aerated soil, the soil will settle and plant roots and crowns can become exposed. Irrigate tilled soil well (or wait for rain) to let settling occur before planting.

Planting in soils with heavy organic matter. Much of the organic matter decomposes within a few years. If a high percentage was used to amend a soil, it is likely the soil will settle, potentially exposing plant roots. If this is the case, plant slightly deeper than normal, but be careful not to cover plant crowns or root flares with soil. Instead, mound up soil around the plant, but far enough away from the plant's center that rain and irrigation water don't wash soil onto the plant crown.

⋏ This soil has been heavily amended with leaf compost and tilled hard. It will settle over time as rain water compresses leaves and organic matter decomposes. Install plants deep enough to prevent exposed roots.

⋏ Should root systems dry out, soaking them before planting helps roots absorb more water than simple irrigation. Completely dry containers are very hard to irrigate, especially if plants are grown in peat-based, hydrophobic soil media.

⋏ Trees and shrubs are often kept in containers longer than perennials and are more likely to be root-bound. Detangle roots carefully, as this step can cause major damage.

⋏ Tools like a Japanese hook can be helpful in detangling roots.

Get rid of "candy" soil. Remove pampering peat-based soil entirely if you are planting large-sized container plants. This step is not necessary if you're using landscape plugs, because they have far lower soil volumes. Before planting, fill a trash can or wheelbarrow full of water and immerse the container plants fully in water, until all air bubbles have come out and the root ball is thoroughly saturated. While the root ball is in water, gently de-tangle roots with a hook, allowing potting soil to fall off.

Never bury root flares and plant crowns. Plant trees and shrubs high enough to allow their root flares to form properly. Prevent untrained maintenance crews from burying the sensitive flares in too much mulch. Only a few floodplain species, such as *Platanus occidentalis*, will tolerate covered root flares without developing diseases or circling roots that will eventually choke a tree. Perennials react even more quickly to having their sensitive crowns covered with soil or mulch. Some species might show signs of

⋏ Only arrange as many plants on site as your team can install before lunch or break time. Avoid leaving plants sitting for too long without irrigation.

∧ Pressing down on soil firmly after planting fills air gaps around roots and allows them to connect with surrounding mineral soil.

⋏ Take great care to avoid installing plants too deeply or not deep enough.

∧ Watering in plants after installation is essential and helps fill underground air gaps with sediment.

A storm water management system in Lancaster, Pennsylvania, ready for planting (top right), and after careful installation (below right).

disease or rot within just a few days if the soil is moist enough. This can be a significant problem if you are planting on steep slopes with loose soil that shifts downward over time. Stabilize slopes first with erosion-control matting or coir logs. Once the soil is securely held in place, the slope should be planted from the top down to avoid covering perennials with too much soil.

Designed plant communities only function if all of their parts are installed correctly and survive the transition from nursery to landscape. Plants are most vulnerable during and immediately after installation. New light levels, more exposure to sun and wind, and new pests and diseases enter their world. In addition to the complexities plants face, designers encounter increased challenges as well. Many parties are involved in the installation process, including clients, nursery personnel, shipping crews, and construction staff. Successful projects require clear communication among all parties and dedication to executing the design vision. Designers must insist on correct procedures and be present to advocate for the design. Only if designers are connected with the work on site will a project transition smoothly to the next step—its long-term care.

Now that the hard installation work is done, the most exciting and enjoyable part begins: watching a unique plant community form and evolve. Plants start to develop and interact, sorting themselves into niches. All of the layers you put in place will become visible as plants mature. Your community will reveal whether your recipe for plant layering is working, and if the selected species were indeed suitable for the site.

CREATIVE MANAGEMENT: KEEPING DESIGNS LEGIBLE AND FUNCTIONAL

Our emphasis here is on creative management, not traditional maintenance. The latter focuses on treating individual plants differently: spraying a fungicide on roses, giving extra water to a hibiscus, pruning a yew to stay under a window. Creative management focuses on gross actions meant to preserve the overall community. This kind of management is guided by goals, which give purpose to actions, as opposed to the blindly applied procedures of traditional maintenance. These goals emanate out of your vision of the archetypal landscape you are trying to evoke, the same objectives and patterns that shaped your design.

Because communities are dynamic, managing them is a creative process. Over time, how a planting is managed has as much, if not more, weight in determining what that planting becomes and how it looks. It is an iterative process of reading the changes in your community and softly massaging things. This process requires adapting not only the planting, but also your strategies and techniques, based on what happens on the ground.

Creative management also elevates the need for designers to collaborate with management crews, explaining the design goals and discussing various techniques for achieving them. The results of unguided management can be disastrous—including a planting that falls into disrepair and eventually disappears. Designers must be part of a planting's life as regular and ongoing consultants.

THE NEED FOR MANAGEMENT

Every planting requires oversight. Even ultra-urban rooftop plantings or streetscape containers engage in natural processes, such as succession, competition, and symbiosis. Spontaneous vegetation has the ability to colonize any area, coming up between young plants and threatening the integrity of the planting. Sites with deep, fertile soils are particularly vulnerable. Tree pits, rain gardens, and cultivated gardens often have rich soils, an ideal habitat for a range of undesirable plants. The speed at which a young planting can unravel and lose its diversity and functionality is alarming, especially in the presence of aggressive clonal species of genera such as *Lythrum* and *Phragmites*. Site managers

continue to be amazed by how quickly rain gardens can turn into a solid monoculture of *Typha*. Without management, a designed plant community can quickly transition to another type of plant community the site supports—a community you and others may not want.

Of course, a range of strategies can be employed to counter the invasion of undesirable plants. But even then, smart management is necessary. Even though we use species

222

naturally supported by a site, and layer plants densely to cover, those tactics don't always prevent spontaneous invaders. Proper species selection for each layer merely reduces the amount of work and resources necessary to keep plantings visually and functionally on track.

Which spontaneous vegetation to keep, what to remove, and how to remove it are the core decisions of planting management. Some voluntary plants are necessary to keep populations of short-lived species alive. Others can help transition one archetype into another. For example, voluntary tree and shrub seedlings can help transition a meadow into a woodland plant community. The loss of certain plants is not always bad, either, particularly when they prove themselves to not be adaptive enough for the site. Some species may decline due to limited planting area, genetic diversity, population size, or disturbance. In the best case scenarios, spontaneously occurring desirable species actually eliminate the need to replace dead plants. They are the single most sustainable planting method ever—the perfect alternative to planting expensive nursery stock and disturbing soil during installation.

A designed meadow community (left) is being overrun by alleopathic Japanese stilt grass (*Microstegium vimineum*). Unmanaged, the planting could soon transition to a *M. vimineum* monoculture (right).

If keeping biodiversity high is a goal, then a form of population enhancement may be necessary to preserve the visual and functional quality of the planting. For example, *Echinacea purpurea* is a highly attractive, long-blooming meadow plant, but often it is short-lived. A series of hot and dry summers can prevent their seed from forming in large quantities, diminishing their numbers in a community. To prevent undesirable species from filling in the gaps, new plants of either the same species or others may need to be added to ensure rich visual and functional diversity. In particular, the filler layer—with its short-lived perennials or biennials that fill gaps in a planting—often contains many species that may need to be replaced over time if they are key elements of the overall design.

MANAGING THE LAYERS

Two overall goals should shape your management practices: preserving the legibility of the design, and making sure the planting still functions in the way it is needed. In an established community, many of the layers bleed together into a cohesive whole. While this may be desirable aesthetically, it can make the specific management activities difficult to understand conceptually. So, because the community was conceived and installed in separate layers, it is helpful to think of management strategies tailored to these various layers. Horticulturally oriented plantings that depend on ornamental effects will have different goals than a more functional storm water planting in a rural area. Remember, context is crucial. Structural layers tend to reside more in the realm of horticulture and ground layers follow ecological land management principles.

Keep orderly frames clean and neat

The frame conveys as much of one's impression of the planting as the planting itself. Clean sidewalks, painted fences, and clipped hedges indicate care. The public may not even be aware of the subconscious message well-kept frames send, but one can clearly read the effects in visitor behavior. Well-tended frames keep trash, dogs, and people out of plantings. They increase the acceptance of more naturalistic design and generate interest in designed plant communities as landscape solutions.

Preserve legibility of the structural layer

Through their height, structural plants create visible patterns, frame views, and define space. Perhaps more than plants in any other layer, exact location, placement, and

Three years after installation, *Eupatorium perfoliatum* forms
a strong seasonal theme in this mesic meadow.

numbers matter. This is the least dynamic layer of all, so management practices should
focus on preserving it essentially as it is. Few spontaneous species appear in exactly the
right place to be incorporated in the structural layer. More often, individuals are lost due
to pests and disease, the natural end of a plant's life, or disturbances such as foot traffic or construction work. Replace lost species, if not with the original plant, then with
another structural plant that serves the same purpose. Replacements are often planted
from containers to allow for exact placement on site. Their quantities and location are
guided by the original planting plan and its bigger design idea.

Keep seasonal themes strong

Seasonal theme plants work as a whole and individual plant placement is less critical.
The number of individuals must be kept high enough that color and texture themes
continue to have impact. Theme plants that spontaneously appear are relatively easy to
incorporate. In fact, plants of this category can be vigorous. They sometimes come up

from seed too thickly, and have to be reduced in number to prevent them from out-competing ground-covering species. Management of this plant element includes selectively removing or weakening such spontaneous additions, by cutting back seed heads or trimming back when they are most vulnerable. This preserves the fine balance between species of a designed plant community.

Orderly frames around this naturalistic design are kept in great shape.

Keep soil densely covered with plants

In the ground layer, various clonal and seed-spreading species mingle with one another. They tend to be highly vigorous and competitive, covering soil and preventing undesired species from gaining a foothold. Nevertheless, various issues can lead to gaps in the ground layer. Management must focus on keeping the soil covered with desired plants, and spaces in the vegetative cover must be recognized and filled as soon as they occur.

If the species composition of this layer is not balanced, a few species can overwhelm other populations and form monocultures over time. Species diversity and ecological value plummet if only a few plants start dominating this layer. Unlike in structural and seasonal theme plant layers, this is the place where diversity can be higher without making a planting look too busy or illegible. For example, various species of sedges and ferns with similar morphology can be mixed in the ground layer without anyone noticing. They may appear like an orderly mass of a single species, but provide all the benefits of diversity.

Do not clean up debris like leaf litter, unless it accumulates too thickly and causes functional problems. A healthy soil maintains its rich microbial life by recycling debris

Replacement plants must be installed soon after plant losses are noticed to keep soil covered, prevent erosion and weed outbreaks, and restore the function of a planting.

In this case, restoring the plant layer also restores the storm water treatment function of a rain garden.

wherever it falls to the ground. Signs of nutrient deficiencies are scarce in dense plantings where debris is allowed to decompose on its own. If the look of leaf litter is problematic due to context or client preference, consider collecting the leaves, then mowing them with a mulching mower. Shredded leaves can then be redistributed as a light mulch layer in the planting.

Evaluate the need for filler plants

As a planting fills in, gaps between individuals become rare and filler plants tend to disappear from planting. Filler species are temporary and allowed to fade as plantings mature. The few years it takes to establish a planting often gives these species enough time to establish a rich seed bank in the soil, which is activated if disturbance occurs.

Filler plants then spring back, until slower growing but more competitive species out-compete them again. So in many ways, a lack of filler species over time is typically a sign of community health, indicating a good ground cover and little open soil.

If for some reason filler plants do not automatically reappear after disturbance, they can either be seeded or planted into open gaps. Maintaining this seed bank of desirable filler plants is important, as this helps communities self-heal when there is a disturbance. Great care must be taken to not disturb a community's other plants in the process.

A MONITORING GUIDE

One of the simplest and most effective tools for designers to use in communicating with management crews and land managers is a monitoring guide. Ideally, this guide, along with the planting plan itself, would be submitted to all available management staff. Designers should meet with management crews just after installation to explain both the design intent and what crews should monitor.

The guide should define long-term management goals, such as keeping the soil covered with desired plants. Guides must be usable in the field and written for management personnel. Correct terminology and language matters. The guides must not only help crews detect problems, but also explain how to handle potential problems. Immediate action is always encouraged through the use of a "toolbox" that gives crews options. After an action is completed, more monitoring will show if the applied tool actually fixed the problem or if other actions are needed.

THE TOOLBOX: MANAGEMENT PRACTICES FOR DESIGNED PLANT COMMUNITIES

Our management toolbox combines traditional horticultural maintenance elements (for example, weeding, watering, and deadheading) with ecological landscape management tools (for example, burning, timed mowing, and enhancement seeding).

All management actions should be as sustainable as possible and avoid unnecessary site disturbance. For instance, soft management techniques such as mowing and selective cutting back are preferred over more energy-intensive weed pulling or herbicide spraying. Filling open gaps within a planting to prevent weeds in the first place is the best solution and requires the least resources in the long run. If weeding is necessary, disturbed soil must be covered with other plants or temporary mulch immediately

MONITOR FOR	CHECK FOR THESE PROBLEMS	YES/ NO	POTENTIAL CAUSES	ACTIONS
Overall aesthetic quality, legibility of design layer patterns, and intactness of orderly frames	Is there a presence of weeds or invasive species?	Yes	Young planting, ground cover not filled yet.	Remove weeds. Strengthen desired species by watering. Add more ground covers if necessary.
			Nearby seed sources.	Remove seed sources.
				Remove weeds and replace with desired species.
			Seasonal gap in plantings.	Contact designer or local plant professional to fill gap with appropriate species. Install ASAP.
		No		Monitor again next time.
	Is the planting legible and aesthetically pleasing?	Yes		Monitor again next time.
		No	The planting is far from original design.	Contact planting design or local plant professional. Develop enhancement planting or editing strategy. Apply strategy. Monitor again.
	Is trash or debris impacting overall appearance?	Yes		Check why it is present. Remove source and debris. Monitor again next time.
		No		Monitor again next time.
Level of biodiversity	Did species disappear since installation?	Yes	Aggressive plants outcompeted less aggressive ones.	Evaluate the plant combinations. Create space and provide resources for less aggressive species to get established.
		No		Monitor again next time.

MONITOR FOR	CHECK FOR THESE PROBLEMS	YES/ NO	POTENTIAL CAUSES	ACTIONS
Functionality	Does planting percolate water?	Yes		Monitor again next time.
		No	Clogged drain.	Clear drain.
			Accumulation of debris.	Remove debris.
			Soil won't percolate.	Contact planting designer or engineer. Retest soil percolation and develop strategy to improve percolation rate.
	Does planting attract pollinators and birds?	Yes		Monitor again next time.
		No	Not enough or right kinds of species to attract insects.	Contact planting designer. Add more desired species in appropriate layers.
			Flowers don't bloom at right time to attract desired species.	Add more appropriate species.
	Does planting control erosions?	Yes		Monitor again next time.
		No	Young roots have not had time to hold soil.	Re-install disturbed plants. Reinforce eroded areas with soil and geotextiles.
			Plants do not have deep enough roots.	Add or replace with appropriate plants.
Density of ground cover	Is bare soil visible?	Yes	Plants died due to drought.	Replace with appropriate species.
			Plants died due to pests.	Identify pest, remove if possible. Replace with resistant species.
			Plants died due to disease.	Identify disease and reason it spread. Replace with resistant species.
			Planting was disturbed.	Identify disturbance and what is necessary to prevent it from happening. Replant as soon as possible.
			No apparent reason.	Contact planting designer or local professional to arrange site visit, identify problem and take action.
		No		Monitor again next time.

MANAGEMENT TOOLS	BENEFITS	CHALLENGES
Burning (selective or large areas)	Controls cool season weeds (such as species of *Alliaria*, *Brassica*, and *Lamium*). Removes dense vegetation and thatch. Strengthens fire-adapted species (such as *Sporobolus heterolepis* and *Schizachyrium scoparium*).	Challenging near infrastructures. Weather dependent. Requires professional supervision.
Cutting back and mowing	Rejuvenates grasses and forbs. Removes dense vegetation and thatch. Manages woody weeds. Controls reseeding of planted species and weeds. Soft management of competition between species. Keeps edges neat around planting .	Cost effective. Easy, does not require expensive equipment. No soil disturbance.
Selective removal of seedlings	Keeps design legible.	Disturbed soil lets weeds come up. Therefore, gaps must be filled immediately with desired plants or seed.
Weeding	Manages spontaneous vegetation. Keeps design legible.	Disturbed soil lets weeds come up. Gaps must be filled immediately with desired plants or seed. Protect desired plants during weed removal.
Spraying (spot-spraying or treatment of larger areas)	Manages undesired vegetation. Manages invasive vegetation. Keeps design legible.	Hard to keep desired species alive if mingled with target species. Requires professional oversight and equipment.

MANAGEMENT TOOLS	BENEFITS	CHALLENGES
Directional pruning	Builds resilient trees and shrubs.	Requires professional oversight and equipment.
Watering	Benefits more moisture loving species. Can shift species composition toward desired aesthetic.	Unsustainable. Costly in some regions.
Fertilizing and amendments	Benefits nutrient-loving species. Can shift species composition toward desired aesthetic.	Unsustainable and costly. Can encourage weeds. Can cause strong vegetative growth at the expense of root development and longevity.
Enhancement planting	Fills gaps to prevent weed outbreaks. Restores legibility of design.	Disturbs soil. Protects established plants during planting process.
Mulching	Protects soil. Suppresses undesired vegetation.	Costly. Can introduces new weed seed. Suppresses seedlings of desired species as well.
Nutrient removal	Can bring elevated nutrient levels to healthy levels. Strengthens plant health and extends lifespan. Less plant vigor can create better aesthetics.	Removing debris is labor intensive. Not all debris is safe to compost. Check for contamination (for example, heavy metals in urban rain gardens). Slow process; results often not visible for decades.

Burning in late winter removes cool season weeds. Here, nutrients stored in the burned biomass of *Lamium purpureum*, *Alliaria petiolata*, and *Allium vineale* are recycled back into the soil.

Spray paint and/or colorful ribbons help train crews to recognize problematic species.

Weeding undesirable species may be necessary but immediate seeding or replanting is recommended to fill left-behind gaps.

to minimize the amount of bare soil. If working with equipment is necessary, use small-sized machinery. For example, when cutting back a planting in late winter, use a small mower or string trimmer to prevent soil compaction and disturbance. Larger tractor-operated mowers can leave ruts in the soil or scour vegetation. Remember, none of these techniques are inherently good or bad. What matters is the context in which they are applied.

SHIFTING MANAGEMENT GOALS OVER THE LIFETIME OF A PLANTING

Three developmental stages can be identified for planting, each with its own goals and monitoring criteria. The first is the PLANT ESTABLISHMENT PHASE, which starts after instillation and, depending on the plant, site, and time of year, can take a few weeks or several months. Once all plants can support themselves, we shift our focus from the individual to the entire planting. This is the LANDSCAPE ESTABLISHMENT PHASE, which includes all the time it takes for a designed planting to fill in and develop its mature shape and forms. By the end of this phase, ground covers thickly carpet the soil, structural species provide frame and order, and seasonal theme plants create spectacular color and texture at various times of the year. Development does not stop here. The final period is the POST-ESTABLISHMENT PHASE, which continues through the life of the planting.

Changes from phase to phase rarely progress evenly within a planting. Some plants or parts of a site may establish faster than others. Disturbance such as a tree falling or a deer eating a patch of ground covers may start over the process of establishment. So in many ways, landscape establishment is more like a gradient between establishment and maturity—with various parts of the planting constantly moving back and forth along this gradient. Always understand what phase each part of your planting is in and use the right goals and management tools.

PLANT ESTABLISHMENT PHASE

At the beginning, the focus is to help each plant survive and get established. This phase starts at installation and ends when all plants are rooted into the surrounding soil and able to support themselves. The development of roots and underground storage organs is the goal of this phase. The main threats that management must address are erosion, plant loss, and weed invasion. The only solution to all of these threats is establishing 100 percent ground cover with desired species more quickly than weeds. In other words, you need every transplant to survive and grow as quickly as possible. Good site preparation and installation practices pay off now. Beyond that, you can help plants get established, but they have to do the growing work themselves. Plant establishment is not about faking happy and healthy plants by artificially triggering lush foliage and flowers.

It is about aiding plants in the process of rooting in and connecting with surrounding soil. Plants put their focus on building an underground foundation first; until that is formed, plants—especially taprooted species like *Asclepias tuberosa* and *Baptisia australis*—will not show much new foliage. It may look like the plants are just sitting there, when in reality they are growing immensely underground. Some clients misinterpret this as a lack of plant vigor and think the solution is applying fertilizer or even replacing plants. Educating a client about the process of plant establishment can prevent common mistakes during this phase and give plants the time they need.

Newly installed plants must be rooted into their new soil by the time winter arrives, or frost heaving may occur. Repeated cycles of freezing and thawing cause water in the soil to expand and contract, pushing up plants and their roots. The crown is then exposed and sensitive roots suffer under cold temperatures and drying winds. Many plants may be seriously damaged or killed, requiring major replanting in the spring.

It is often beneficial to cut back plants to prevent flowers and seeds from forming. This may reduce the showiness of a planting in its early life, however, more plant energy is channeled into developing a strong crown and root system. Cutting back encourages many perennials to sprout new foliage from the base, making plants thicker and stronger. Flowers and seed come later. Temporary filler plants such as annuals and biennials can be used to bridge this time visually and provide early interest in the form of flowers and texture. This can be accomplished by overseeding.

Keeping deer, rabbits, and other herbivores out of the planting is essential during this phase. Young plants are especially vulnerable. Plants are still adjusting to UV light and their leaves are not yet hardened off, making their tender foliage especially tasty to herbivores and insects. Temporary repellents can be helpful until plants are hardened off and less palatable to wildlife. If pest pressure is high and threatens the entire community, designers should consult a specialist and apply appropriate measures.

If plants die during the establishment phase, pay careful attention to why they failed. Some plants die because they do not fit the conditions of a site. Even the most careful site analysis and researched plant list cannot predict how plants will perform on a site. Noting which plants thrive and which ones languish provides feedback about what wants to grow on the site. If a plant deteriorates, do not always assume that poor selection is the culprit. Different plants establish under different conditions, so the

Management Tip
—

Take care not to cut too low and hurt the crown of a plant or remove too many leaves before it has formed large enough underground storage organs to cope with heavy loss of biomass. Some species may not respond well to cutting back at certain times of year. For example, if evergreen grasses like some species of *Carex* are cut back too low after they flower in early spring, they will often suffer heavy losses during late frosts because they do not have enough biomass to protect sensitive plant crowns and recover from cutting. If in doubt, reach out to a horticultural professional for guidance.

At the end of the plant establishment phase, all transplants are fully rooted in and can support themselves with water and nutrients.

timing or weather during the first few weeks may favor certain species and discourage others. A cool, wet spring may favor aggressive, clonal, spreading genera like *Monarda* or cool season growers like *Festuca*, yet discourage warm season grasses like species of *Sporobolus* or *Eragrostis*. The latter may still have been good selections for the site, but were thwarted by weather. Try to understand why plants perform differently; the information will provide valuable clues about the site and help direct enhancement planting if needed.

LANDSCAPE ESTABLISHMENT PHASE

Once all plants are established, the planting enters its next phase: growth and development. With every inch plants grow, the community's layers and design are revealed. Structural species gain height, and ground covers rush to close any bare soil. Seasonal color and texture themes become visible, at first faint but increasingly stronger. Not all species develop uniformly; some may reach mature size long before others. Slowly and inconsistently, the design emerges.

But as the design takes shape, mistakes are quickly evident. This is the ideal time to address them. Problems may include unsuitable species for structural frames. For example, if a structural plant frequently flops after rain, it should be replaced. If a ground cover grows spotty, you may want to add more competitive species to the mix. During this phase, plants will start to grow into other plants' territory. Plant dynamics and competition starts. Some species may be bested; others may happily expand their population. Temporary cover crops and short-lived filler species generally decline as more aggressive species come to dominance during this phase. Species that fail now are most likely not fit for a site, assuming they were properly installed.

The conclusion of the first full year of growth is often a good time to reevaluate a site and make adjustments, particularly to plants in the design layer. At this point, the planting is grown-in enough to evaluate its design strengths and weaknesses, but not so established that moving plants is too difficult. For example, if your structural layer is not strong enough, new additions may be necessary or existing plants may need to be moved. If your thematic layer is not bold enough, consider adding more plants. If your layers bleed together too much, move some plants around to strengthen patterns. Just remember that any transplanting and replanting starts the establishment phase over, so tailor your management to this.

At the end of this phase a planting feels stable and balanced. All of its parts are in place and interacting with one another. The soil is densely covered with multiple layers of plants and evocative patterns, and seasonal themes capture our attention.

Management Tip
—

Make sure to set aside resources to cover necessary enhancement planting. Rarely does initial installation result in 100 percent success. There are almost always species that fail and gaps that need to be filled once plants are established. Let your client know in advance that the planting will be two-phased, with an initial and an enhancement planting. By planning for these phases and building them into your cost schedule, the enhancement is not perceived as "extra" but merely part of installation.

BEYOND ESTABLISHMENT: CREATIVE MANAGEMENT FOR THE LONG TERM

Plant communities constantly change. Some of this change is slow and subtle, such as a sedge gradually gaining dominance along a slightly wetter fold in a field, or the thinning out of grasses as a tree's canopy creates more shade. Other changes happen faster, such as a large storm toppling a stand of white pines that served as a backdrop to your planting, or an invasion of *Phragmites*. Change is unavoidable.

Designed plantings generally will not last without continued management. Decades of experience from designers all over the world has led to the same conclusion: there is no naturally stable plant list, and stability is not possible without management. In fact, if left alone, even the most elegant and thought-through plan will evolve into something completely different. The question for the designer or land manager is, how much change do we allow?

Cutting herbaceous plants back after the winter rejuvenates them and keeps woody seedlings from establishing. Do not cut back too hard and expose soil—an invitation for weeds. Here, dead debris has been removed after cutting plants high.

<< Annual species should be selectively cut back to prevent soil disturbance and reseeding.

< Cut back large, undesired plants at the base instead of pulling, which can harm root systems of surrounding species and disturb soil.

For most plantings, the goal is not to avoid change; the goal is to achieve some level of stability through management. To do this, management generally focuses on coarse actions applied to the overall community. For grassland or some woodland communities, this may mean an annual mowing or burning to keep woody species from dominating or to encourage fire-dependent plants. For some shrubland and woodland communities, including edges of forests, various kinds of thinning can preserve the variety of tree and shrub combinations. Thinning can be especially important to help slow-growing trees like oaks compete against faster-growing shrubs like species of *Rhus* or *Alnus*.

Over time, plant communities can either become more visually dense or more open. The degree of openness or density is a critical design decision that can only be addressed through management. Large sites with a mosaic of different glades, meadows, fingers of shrubland, and stands of trees need management to maintain different levels of openness. Coppicing is a technique in which trees and shrubs are cut down to the ground. Coppicing dense aggregations of shrubs can thicken and rejuvenate these mixes, making them more legible as a design feature. Or in urban areas, coppicing may help to reduce the height and increase the density of a naturalistic hedgerow. British researcher and plantsman Nigel Dunnett has long been an advocate of coppicing as a creative tool for managing mixed woody vegetation on public land. Dunnett points out that coppicing can reduce shade and create a richer mosaic of woody and herbaceous vegetation. Coppicing is typically done on five- to ten-year rotations, allowing trees and shrubs to recover between cuttings.

In other contexts, designers may decide to not coppice mixed plantings with multistemmed trees and high shrubs. Coppicing these plants will make them dense and impenetrable, but allowing them to grow over time into low woodlands can create spaces that people can enter and inhabit. These woodland rooms can be very pleasant, particularly if peninsulas of woodland are set against more open, grassland-type vegetation, allowing sheltered views into wide open expanses.

Separating a planting into different zones is one way to identify priorities and make the most of a management budget. Zones may be arranged as a gradient, with plantings closer to pedestrian paths and buildings receiving higher input, and plantings farther away receiving less. The *heemparks* in the Amsterdam neighborhood of Amstelveen use this approach to great effect. These parks include a series of public plantings designed in the mid-twentieth century to thrive in the acidic, wet soil of the site. The plantings continue to thrive through creative management practices. Some plantings are managed less intensively, such as wildflower meadows and woodland edge plantings along roadsides. In areas of higher visibility, a more concentrated stylistic planting is maintained. Large blocks of shady ground covers are punctuated with ferns and perennials; in sunny areas, colorful perennial meadows bloom along paths.

Other long-term management considerations include regular monitoring for invasive species. Remove weeds and invasive species immediately, as these plants can establish populations quickly. Completely remove all parts of these plants, including underground storage organs, with a long blade or other weeding tool. Replant the area with desirable species and monitor regularly.

Long-term management might allow a planting to transition into a different plant community over time. For example, an open meadow might evolve into an open shrubland, or an open shrubland into an open forest. Monitoring and management actions have to follow this long-term goal in order to guide a plant community's development after it is established.

MANAGEMENT AS A CREATIVE DIALOGUE

Design remains essential in all aspects of management. In fact, smart management is a creative process requiring a large vision as well as attention to details. Too few designers stick with projects for more than a few years after installation, but active engagement with clients and land managers can not only benefit the planting, but—if negotiated well—can be a financial benefit to the designer. Management goals can and should shift through a project, so continued conversation with landowners and management crews can make all the difference. Designers, landowners, and managers should connect early after installation to discuss the planting's needs and priorities.

Designers should take a leading role in preparing management schedules and guides, and submit them with the planting plans. Schedules and guides should not be written in the bureaucratic language of typical construction specifications, but should be distilled, action-oriented charts or checklists. Periodic on-site meetings are important to explain how prescribed actions should be translated into reality, and to tweak guides and schedules based on site realities.

Equally crucial to schedules and guides are ample management funds. Successful management depends upon regular, informed action—none of which is possible without a budget. Limit the size and scope of a planting only to that which can be managed and budgeted; failed projects do a disservice to both designers and clients.

Ultimately, management is a variety of relationships—a mental relationship between an idea and a place, a physical relationship between a manager and a piece of land, and even an emotional relationship between our desire for natural beauty and our encounter with living plants. But all good relationships require presence, commitment, and an open sharing of ideas. The best planting projects can do just that. They engage designer, owner, and manager in a dynamic, rewarding connection with each other and a site.

CONCLUSION

Tomorrow's designed landscape will be many things—more plant driven, site responsive, and interrelated—but one thing it will not be is stylistically the same as its predecessors. It is perhaps easy to assume that plantings layered with a diverse mix of species would be necessarily naturalistic in style. In many cases, this is true. But gardens of any style can benefit from applying natural principles. Whether the planting is formal or informal, classical or modern, highly stylized or naturalistic does not matter. What matters is that plants are allowed to interact with other plants and respond to a site. This is the essence of resilient planting.

We want to dwell briefly on three designed plant communities—one formal, one soulful, and one playful—to highlight the different expressions plantings can take.

A MEDITATION ON THREE GARDENS

HEINER LUZ'S FORMAL RESIDENTIAL GARDEN

The client, an architect, bought a historic residence in Munich, Germany, and contacted Heiner Luz to help renovate the garden. The client had several specific requests. The landscape had to match the formality of the building. Parts of the garden are protected by historic restrictions and needed to be protected and restored, however, the client wanted to include contemporary perennial planting as a part of the garden. Luz developed an elegant solution: using clipped boxwood parterres, he framed a series of richly layered perennial plantings. The boxwood frames give the garden its structure and relate to its historic context. Multi-stemmed *Koelreuteria paniculata* punctuates the center of each bed. Yet within the beds, a mixed planting layered like a natural community provides interest throughout the year.

Seasonal themes are formed by perennials and bulbs. The color palette for the planting is limited to yellow and white in order to keep the look formal and elegant. The overall character of the planting is very cheerful, even though few colors are used. Seasonal themes shift from predominantly white at certain times of year to mainly yellow during others. In February, a mix of *Eranthis hyemalis* and *Galanthus nivalis* start the flower season. They are followed by yellow *Narcissus* 'Hoopoe' in April. Then, *Hemerocallis* 'Stella d'Oro' flowers provide yellow from May to September. In some areas, species of *Hemerocallis* are combined with *Myrris odorata*, *Eurybia divaricata*, and *Cimicifuga simplex* 'Armleuchter'. Ground is covered by *Alchemilla epipsila* and *Phlox divaricata* 'May Breeze'.

This garden combines formality with lively, cheerful planting in a brilliant way. It proves that a layered, biodiverse garden does not have to look like a sprawling meadow. The palette used in this project is cosmopolitan, but the same design could have been created with native species. The garden is a testament to the stylistic and artistic adaptability of community-based planting.

Formal hedges surround a complex plant community in this historic setting in Munich, Germany (below). The garden, designed by Heiner Luz, illustrates how designed plant communities can fit into even restricted spaces (bottom).

JAMES GOLDEN'S FEDERAL TWIST

The garden of James Golden near Stockton, New Jersey—named Federal Twist for the road it borders—began as an unlikely experiment to create an artificial wet prairie in a terribly inhospitable site. Golden and his husband Phillip purchased a mid-century house on four acres of woodland, a weekend retreat from their home in Brooklyn. The house sits atop an elevated mound, looking over what had been an open field surrounded on all sides by over 500 acres of preserved woods. The framed views into the field inspired the garden; here would be a prairie.

Left alone, the site was returning to forest. But the openness beckoned to Golden. To make room for the garden, he felled seventy cedars, using the wood chips to make a series of serpentine paths that form the bones of the garden. Beyond the clearing, the rest of the garden's creation was a kind of radical acceptance of the given conditions. Those conditions were far from ideal—standing water in many parts of the garden; heavy, wet clay, particularly in winter; tall trees that limit light into the garden; and cold winters that delay spring growth. Any one of those conditions might have driven other gardeners to more drastic measures. But Golden used them as a starting point. He did not prepare the soil, spray herbicide to get rid of weeds, or mulch. In fact, Golden planted directly into the existing bed of weeds, utilizing the stability of the green cover to establish the taller structural plants. This approach allowed Golden to discover quite a variety of desirable plants, including many useful species of *Carex, Scirpus, Juncus, Sisyrinchium,* and *Erigeron* that could have been lost if he had used a mass herbicide for weed removal.

The ground cover layer has evolved with the garden. Faced with the omnipresence of weeds and invasive plants like Japanese stilt grass, Golden's strategy was unconventional. Instead of waging ordinary warfare on invasive species, he employed guerilla tactics, striking at vulnerable moments and inserting an array of equally thuggish beneficial plants. He creatively used timing, management, and coarse techniques like seeding, mowing, and burning to shift the balance to plants he wanted in the garden. As a result, the ruderal weeds that started the garden have yielded to more stable mixes of long-lived competitors. In late winter, Golden burns the garden, a process he calls "a purifying ritual of fire and destruction," which leaves the ground stripped to brown earth. But by late spring, the ground is cloaked with a gaudy quilt of contrasting textures. *Equisetum, Onoclea sensibilis, Iris versicolor, Packera aurea, Petasites* and others form a densely woven tapestry. The garden swells in a frothy sea of greens, a relatively restrained moment before the garden bolts upward with the summer heat.

With the heat comes the real show. Federal Twist's tour de force is its collection of tall, upright plants. Wet prairie stalwarts like *Symphyotrichum, Filipendula, Rudbeckia maxima, Eutrochium, Vernonia,* and *Silphium* mix with exotic imports such as *Miscanthus,*

The garden swells in summer, as structural perennials grow over ground cover species.

Sanguisorba, and *Inula*. In spring, the long, open views across the garden give the visitor a feeling of expansiveness; by late summer, the visitor shrinks beneath the towering prairie, absorbed in the sublime foliage. Perhaps the garden's finest moments are in fall and early winter, when the vast structural perennials turn skeletal with early frosts. Slanting light through the woodland hits the prairie with dramatic force, turning grasses into glowing embers.

Red-painted stumps provide an orderly frame for a quieter corner of the garden.

Golden's garden is a daring dance with nature. It is a garden of contrasts: always pushing the line between control and chaos, artifice and naturalness, darkness and light. Part of its enduring appeal is the way it challenges one's perception of what a garden is. Because the planting is such an integrated community, all responding to a specific place, Federal Twist artfully confuses the distinction between native and exotic, making harmonious combinations out of Himalayan daisies and New England asters. The entire garden flows out of a profound acceptance of what exists as a way of creating something utterly new and expressive. In this way, Golden uses the garden, with all of its changing objects, as a foil to explore the quieter, emotional undercurrent that drives him to create.

DEREK JARMAN'S PROSPECT COTTAGE AT DUNGENESS

Derek Jarman was a hugely influential British filmmaker and writer. Toward the end of his life, he created Prospect Cottage, a simple wood house that stood on the shingle beach of southwest England. The cottage is one of several fishermen's shacks, wedged on the beach between the English Channel and the Dungeness nuclear power plant. It is a brutal landscape. Nature is overwhelming: sun, wind, and sea salt continuously scald the beach. The horizon stretches in all directions, only interrupted by power poles or the flashing lights of the power plant. Yet within the sunbaked shingles, a garden grows. Sea kale and poppies bloom among the flotsam that Derek arranged throughout the garden.

The site has a kind of post-apocalyptic feel to it. The pebble desert, the leftover ruins of seaside industry, and eerie lights of the nuclear plant all allude to a kind of dystopic, *Mad Max* landscape. Jarman's own arrangements heighten this effect; pieces of black flint are set around the garden like miniature dolmens. Yet amid this barrenness, a designed plant community flourishes. Jarman's process of building the garden was additive. At first, he did not think the planting would survive at all. But early success with a dog rose encouraged more experimentation. Soon species of *Santolina*, sea kale, and valerian were added among the rubble. Some of the planting is arranged in formal rectilinear beds, other spaces float like little islands in the shingle. Jarman discovered that the soilless beach actually supported a wide range of plants. The gravel is an ideal medium for germinating seed—warmed by the sun on top, but cool and moist below. A whole range of self-seeding plants, including viper's bugloss, foxglove, and campion found a ready home in the garden. Annuals play an important role in the garden. Casually strewn poppies, marigolds, and species of *Helichrysum* create strong thematic bursts in the summer, providing a splash of color that is one of the few distinctions from a shoreline bleached of color.

Prospect Cottage's poignancy is sharpened by the fact that Jarman created it while dying of AIDS. Jarman's imminent death did not stop his act of creation, but instead infused it with new vitality. The plants' struggle to live in the windswept gravel mirrored Jarman's own struggle. In the end, the garden endures, offering a living testament of the irrepressibility of love amidst the cruelty and indifference of nature.

As a result, the garden has achieved a kind of cult status among gardeners, artists, and plant enthusiasts. Part of its enduring appeal is that the planting so seamlessly merges with the place. It is a garden without boundaries. What defines the "inside" of the garden is that it is simply a more intensified, stylized

Colorful flowers such as species of *Santolina* and *Eschscholzia* grow right out of the gravel and contrast dramatically with the dark cottage in the background. No site seems to be too harsh for plant communities to thrive.

version of the wild plant communities that surround it. Jarman's own love of the existing landscape sensitized him to seeing plants that would accentuate the desolate beauty of the place. Daffodils, he wrote, looked silly on the beach; yet so many other selections amplified the seaside setting.

What is perhaps most captivating about the garden is its playfulness, particularly in a serious setting. The planting serves many functional purposes: ornament, a nectar source for his honey bees, and an ecological solution for a harsh site. Yet there is absolutely nothing serious about the garden. Jarman played in the pebbles, assembling found treasures into little dioramas. Even his plant selections favored cheerful marigolds and poppies over more dour native perennials. Perhaps this is what gives the garden its edge. On the one hand, it surrenders to the unavoidable brutality of the site; yet it responds defiantly by blooming and flourishing. The garden continues to be celebrated today, a tribute to its timelessness.

. . .

These three gardens show the wide range of contemporary expressions that community-based planting can have. Each accepts the givens of its site, yet transcends its limitations, resulting in a distilled, emotionally accessible vision of the place. Together, they span the gradients of formal and informal, art and ecology, and serious and playful—showing that this method can indeed be applied to almost any kind of planting.

As populations expand and resources become increasingly limited, plantings can no longer be just ornamental backdrops for our buildings. They must instead perform double duty: cleaning our storm water, providing a food source for pollinators, and acting as a kind of genetic reservoir for diversity. Achieving this requires understanding how plants fit together, how they change over time, and how they form stable compositions.

WHY DESIGNED PLANT COMMUNITIES? WHY NOW?

    A community-based approach provides a method for more functional planting. It addresses the single biggest factor for instability in landscape: bare ground. It focuses on achieving a rich density of species in lower layers of the planting that allows for design flexibility in the upper layers. Significantly, designed plant communities have mechanisms for self-correction and healing, making them more resilient than conventional garden styles. Designers no longer have to predict every scenario. They can instead rely on the community to adapt to changing circumstances.

    Designed plant communities emphasize function, yes, but what we ultimately need are plantings that are relatable to humans. For us, it is their aesthetic and evocative qualities, perhaps even more than their utility, that makes them relevant and timely. Designed plant communities have the potential to transcend many of the bad stereotypes associated with ecological planting. The lingering impression that native and ecological planting is messy partly explains why so much of the world—particularly the United States—relies on lawns and conventional horticulture as the default treatment, despite the high labor and cost needed to perpetuate them. But this stigma of messiness need not be perpetuated. In many ways, a community-based approach to planting depends even more on a designer to translate natural patterns into an ordered vernacular that connects with people.

    This is precisely why a focus on designed plant communities can lead to a renaissance of design. Armed with a basic knowledge of how plants behave ecologically, designers can elevate their work, creating effects not possible with conventional planting.

Successfully layering plants of different competitive types opens the door to new combinations, new styles, and new expressions. Think about it: many of the great innovations in planting design over the last century were influenced by designers studying wild planting. Beth Chatto, in creating her widely celebrated garden in Britain, based much of the planting's cutting-edge style and many of the combinations on the study of plant communities. Likewise, the work of German practitioners over the last four decades—largely reliant on plant communities as models—has resulted in some of the most influential planting in Europe. Even today, the work of designers such as Piet Oudolf, Dan Pearson, James Hitchmough, Nigel Dunnett, Sarah Price, Cassian Schmidt, Petra Pelz, Roy Diblik, and Lauren Springer Ogden (to name only a few) draw upon plants in the wild as sources of great inspiration.

The time is right for a renaissance of horticulture. Designed plant communities require an ecological understanding of plants, but even more, they need designers with an eye for combinations, a flare for color, and an intuitive sense of natural harmony. They need gardeners who can find a place to plant, even among skyscrapers and row houses. They need plant lovers who understand that we don't need to go to a national park to have a spiritual experience of nature; we can have such experiences in our backyards, parks, and rooftops.

If it is true that the next renaissance of human culture will be the reconstruction of the natural world in our cities and suburbs, then it will be designers, not the politicians, who will lead this revolution. And plants will be at the center of it all.

A rich fabric of grasses intermingled with forbs recalls a large grassland, allowing city folk to be immersed in wildness in the middle of an entirely artificial landscape.

ACKNOWLEDGMENTS

In many ways, this book is an effort to pull together the combined thoughts of so many brilliant designers and plants people all over the world. We are enormously indebted to several colleagues and mentors who inspired this idea. To Darrel Morrison who taught us to see the incredible idea that a native plant community represented. To designers Heiner Luz, Sarah Price, Nigel Dunnett, and James Hitchmough, whose body of work is the single best selling point for this book. To writers and practitioners Noel Kingsbury, Norbert Kühn, Joan Iverson Nassauer, Cassian Schmidt, Ed Snodgrass, Lauren Springer Ogden, and Peter Del Tredici, whose thoughtful yet provocative spirits were the undercurrent for this book. And to the late Wolfgang Oehme, who served as mentor and friend to both of us. He would have thought the idea for this book was total crap.

An evocation of the British countryside, designed by Sarah Price.

We are indebted to the many people who made their photos and designs available to us, and would like to thank designers James Golden; HM White; Heiner Luz; Oehme, van Sweden & Associates; Pashek Associates; Sarah Price; and Adam Woodruff for sharing their innovative projects. We would like to thank several talented photographers whose stunning photos say what we cannot in words: Mark Baldwin, Alan Cressler, Hank Davis, Uli Lorimer, John Roger Palmour, Tom Potterfield, Jonas Reif, Elliot Rhodeside, Bill Swindaman, Ivo Vermeulen, and the team at North Creek Nurseries.

We would like to particularly thank the talented editorial staff at Timber Press. Juree Sondker and Julie Talbot guided us through this process with warmth and humor and provided critical guidance as we floundered along the way. Andrew Beckman helped us sharpen the message and make the language more direct. Thanks to other editors and designers as well.

Finally, we would like to thank our families. To Gail and Lee Warner, who provided home-cooked meals and supplemental child care during this process. To Sam and Linda Rainer who taught Thomas everything he knows about writing. To Jude Rainer, who sees the natural world through fresh eyes. To Hendrik and Marion Pfeifer, who provided vital support from a long distance away. And to our spouses, Melissa Rainer and Jim West. This book was written on nights and weekends during their time. They not only endured this long process with grace and support, they continue to stick with two plant-obsessed geeks.

BIBLIOGRAPHY

Beck, Travis. 2013. *Principles of Ecological Landscape Design*. Washington, Covelo, London: Island Press.

Del Tredeci, Peter. Spring/Summer 2004. Neocreationism and the illusion of ecological restoration. *Harvard Design Magazine* 20.

Eissenstat, D.M. and R.D. Yanai. 1997. The ecology of root lifespan. *Advances in Ecological Research* 27: 2–59.

Grime, J. Philip and Simon Pierce. 2012. *The Evolutionary Strategies that Shape Ecosystems*. Chichester, UK: Wiley-Blackwell.

Hansen, Richard and Friedrich Stahl. 1993. *Perennials and Their Garden Habitats*. 4th ed. Portland, Oregon: Timber Press.

Jaffe, Eric. 2010. This side of paradise: why the human mind needs nature. *Observer* 23, no. 5.

Kingsbury, Noel. Clump or mingle? http://thinkingardens.co.uk/articles/clump-or-mingle-by-noel-kingsbury/.

Kingsbury, Noel and Piet Oudolf. 2013. *Planting: A New Perspective*. Portland, Oregon: Timber Press.

Kühn, Norbert. 2011. *Neue Staudenverwendung*. Stuttgart (Hohenheim): Eugen Ulmer KG.

Nassauer, Joan Iverson. 1995. Messy ecosystems, orderly frames. *Landscape Journal* 14, no. 2: 161–170.

Schwartz, Judith D. 2014. Soil as carbon storehouse: new weapon in climate fight? *Yale Environment* 360.http://e360.yale.edu/feature/soil_as_carbon_storehouse_new_weapon_in_climate_fight/2744/.

Seabrook, Charles. June 5, 2012. Tallgrass prairies extend into Georgia. *Atlanta Journal-Constitution*.

Watson, Todd W. 2005. Influence of tree size on transplant establishment and growth. *HortTechnology* 15(1).

Whittaker, Robert H. 1975. *Communities and Ecosystems*. 2nd ed. New York: MacMillan Publishing Co., Inc.

METRIC CONVERSIONS

INCHES		CM
¼		0.6
½		1.3
¾		1.9
1		2.5
2		5.1
3		7.6
4		10
5		13
6		15
7		18
8		20
9		23
10		25
20		51
30		76
40		100
50		130
60		150
70		180
80		200
90		230
100		250

FEET		M
1		0.3
2		0.6
3		0.9
4		1.2
5		1.5
6		1.8
7		2.1
8		2.4
9		2.7
10		3
20		6
30		9
40		12
50		15
60		18
70		21
80		24
90		27
100		30

TEMPERATURES

$$°C = \frac{5}{9} \times (°F - 32)$$

$$°F = \left(\frac{9}{5} \times °C\right) + 32$$

Andrea Cochran Landscape Architecture, Children's Museum, page 228

Andropogon Associates, pages 225, 226

Claudia West LLC, pages 204, 220, 237, 238, 239

Roy Diblik, Shedd Aquarium, page 28

James Golden, pages 60, 122, 166; Federal Twist garden, pages 246–247, 248, 249

HMWhite Site Architects, New York Times headquarters, pages 130, 132–133, 210; Manhattan rooftop meadow, pages 2, 25, 144, 255

Derek Jarman, pages 251, 252

Longwood Gardens, meadow garden, page 83

Heiner Luz, Ziegeleipark, page 171; Munich garden, page 245

Michael Van Valkenburgh Associates and Ed Snodgrass of Green Roof Plants, ASLA headquarters rooftop garden, page 53

Mount Cuba Center, pages 50, 51, 112

Oehme van Sweden & Associates in collaboration with the horticultural staff at the New York Botanical Garden, Native Plant Garden, pages 59, 88, 140, 142, 143, 146, 203

Piet Oudolf, Trentham Estate, page 39

Pashek Associates, David Lawrence Convention Center, pages 24, 57, 188

Sarah Price, 2012 Telegraph Garden, pages 68, 242, 257; Great British Garden, Queen Elizabeth Olympic Park, pages 4, 174

Sarah Price and Nigel Dunnett, European Garden, Queen Elizabeth Olympic Park, pages 26–27, 83, 147

Sarah Price and James Hitchmough, North American Garden, Queen Elizabeth Olympic Park, cover, pages 40, 190; Southern Hemisphere Garden, Queen Elizabeth Olympic Park, page 222

Thomas Rainer, pages 10, 136, 150, 155

Charlotte Rowe, ABF The Soldiers' Charity's Garden, Chelsea Flower Show, page 120

Cassian Schmidt, Hermannshof, pages 164–165

Schulenberg Prairie, Morton Arboretum, pages 12, 78

Lianne Siergrassen, page 39

Terry Guen Design Associates, pages 144, 158, 159

Adam Woodruff, pages 21, 67, 138–139, 151, 156, 157, 208–209

Thomas Rainer is a registered landscape architect, teacher, and writer living in Arlington, Virginia. Thomas is a passionate advocate for an ecologically expressive design aesthetic that does not imitate nature, but interprets it. His planting designs focus on creating a modern expression of the ground plane with a largely native palette of perennials and grasses. Thomas has designed landscapes for the U.S. Capitol grounds, the Martin Luther King, Jr. Memorial, and The New York Botanical Garden, as well as over 100 gardens from Maine to Florida.

Thomas has worked for the firm Oehme, van Sweden & Associates and currently is an associate principal for the firm Rhodeside & Harwell. He has a broad range of experience in project types, ranging from intimate residential gardens to expansive estates, rooftop gardens, botanical gardens, public display gardens, large-scale ecological restorations, and national memorials. His work has been featured in numerous publications, including *The New York Times, Landscape Architecture Magazine, Home + Design, New England Home, Maine Home + Design*, and the *Hill Rag*.

Thomas is a specialist in applying innovative planting concepts to create low-input, dynamic, colorful, and ecologically functional designed landscapes. His work aims at moving planting design from a largely decorative role to one that is essential to addressing the environmental challenges of our day. He teaches planting design for the George Washington University Landscape Design program and regularly speaks to audiences throughout the East Coast on sustainable planting design.

He blogs regularly at the award-winning site *Grounded Design*.

Claudia West is a landscape designer, lecturer, and consultant based in White Hall, Maryland. In her current role as ecological sales manager at North Creek Nurseries, Claudia bridges the gap between project designers, plant growers, installation and management professionals, and ecology. She works closely with design and restoration professionals, offering consultation services from initial project planning stages and site-appropriate planting design, to adaptive management strategies after project completion. Claudia's work is centered on the development of ecologically sound, highly functional, and aesthetically pleasing planting design that stands the test of time. In collaboration with the team at North Creek Nurseries, Claudia developed the first plant community–based design tool and established numerous test and evaluation plantings to further evolve the concept of mixed planting design in the United States.

Claudia learned the principles of plant propagation and stable planting design while growing up in her family's landscape design, plant nursery, and florist business. She served as a design consultant for Wolfgang Oehme/Carol Oppenheimer and was employed at Bluemount Nurseries, Sylva Native Nursery, and Envirens, Inc.

Claudia is a sought-after speaker on topics such as designing and managing plant communities, functional planting for green infrastructure, the application of natural color theories to planting design, and sustainable practices in plant propagation.

Frontispiece: Site-appropriate vegetation densely covers every inch of soil, creating a highly functional, emotionally accessible, and resilient plant community.

Published in 2015 by Timber Press, Inc.

Photo and illustration credits appear on page 260

The Haseltine Building
133 S.W. Second Avenue, Suite 450
Portland, Oregon 97204-3527
timberpress.com

Printed in 2015
Text and cover design by Debbie Berne Design

Library of Congress Cataloging-in-Publication Data

Rainer, Thomas (Landscape architect), author.
   Planting in a post-wild world: designing plant communities that evoke nature/Thomas Rainer and Claudia West.—First edition.
      pages cm
   Includes bibliographical references and index.
   ISBN 978-1-60469-553-3
   1. Gardens—Design. 2. Plant communities. I. West, Claudia, author. II. Title.
   SB472.45.R35 2015
   712—dc23   2015019338

A catalog record for this book is also available from the British Library.